Lean Six Sigma for Optimal System Performance in Manufacturing and Service Organizations:

Emerging Research and Opportunities

Edem G. Tetteh
Rowan College at Burlington County, USA

Hans Chapman
Cape Fear Community College, USA

A volume in the Advances in Logistics, Operations, and Management Science (ALOMS) Book Series

IGI Global
DISSEMINATOR OF KNOWLEDGE

Published in the United States of America by
IGI Global
Business Science Reference (an imprint of IGI Global)
701 E. Chocolate Avenue
Hershey PA, USA 17033
Tel: 717-533-8845
Fax: 717-533-8661
E-mail: cust@igi-global.com
Web site: http://www.igi-global.com

Library of Congress Cataloging-in-Publication Data

Names: Tetteh, Edem, 1972- author. | Chapman, Hans, 1965- author.
Title: Lean six sigma for optimal system performance in manufacturing and service organizations : emerging research and opportunities / by Edem G. Tetteh and Hans Chapman.
Description: Hershey : Business Science Reference, [2018]
Identifiers: LCCN 2017027632| ISBN 9781522540625 (hardcover) | ISBN 9781522540632 (ebook)
Subjects: LCSH: Six sigma (Quality control standard) | Lean manufacturing.
Classification: LCC TS156.17.S59 T48 2018 | DDC 658.5/620218--dc23 LC record available at https://lccn.loc.gov/2017027632

This book is published in the IGI Global book series Advances in Logistics, Operations, and Management Science (ALOMS) (ISSN: 2327-350X; eISSN: 2327-3518)

British Cataloguing in Publication Data
A Cataloguing in Publication record for this book is available from the British Library.

Advances in Logistics, Operations, and Management Science (ALOMS) Book Series

ISSN:2327-350X
EISSN:2327-3518

MISSION

Operations research and management science continue to influence business processes, administration, and management information systems, particularly in covering the application methods for decision-making processes. New case studies and applications on management science, operations management, social sciences, and other behavioral sciences have been incorporated into business and organizations real-world objectives.

The **Advances in Logistics, Operations, and Management Science** (ALOMS) Book Series provides a collection of reference publications on the current trends, applications, theories, and practices in the management science field. Providing relevant and current research, this series and its individual publications would be useful for academics, researchers, scholars, and practitioners interested in improving decision making models and business functions.

COVERAGE

- Operations management
- Services management
- Computing and information technologies
- Information management
- Organizational behavior
- Marketing engineering
- Political science
- Production management
- Decision analysis and decision support
- Risk Management

IGI Global is currently accepting manuscripts for publication within this series. To submit a proposal for a volume in this series, please contact our Acquisition Editors at Acquisitions@igi-global.com or visit: http://www.igi-global.com/publish/.

The Advances in Logistics, Operations, and Management Science (ALOMS) Book Series (ISSN 2327-350X) is published by IGI Global, 701 E. Chocolate Avenue, Hershey, PA 17033-1240, USA, www.igi-global.com. This series is composed of titles available for purchase individually; each title is edited to be contextually exclusive from any other title within the series. For pricing and ordering information please visit http://www.igi-global.com/book-series/advances-logistics-operations-management-science/37170. Postmaster: Send all address changes to above address.

Titles in this Series

For a list of additional titles in this series, please visit:
https://www.igi-global.com/book-series/advances-logistics-operations-management-science/37170

Handbook of Research on Promoting Business Process Improvement ...
Nita H. Shah (Gujarat University, India) and Mandeep Mittal (Amity School of Engineering and Technology, India)
Business Science Reference • ©2018 • 450pp • H/C (ISBN: 9781522532323) • US $345.00

Supply Chain Management Strategies and Risk Assessment in Retail Environments
Akhilesh Kumar (Indian Institute of Technology Kharagpur, India) and Swapnil Saurav (JDA Software, India)
Business Science Reference • ©2018 • 351pp • H/C (ISBN: 9781522530565) • US $225.00

Management Control Systems in Complex Settings Emerging Research and Opportunities
Filippo Zanin (University of Udine, Italy) Eugenio Comuzzi (University of Udine, Italy) and Antonio Costantini (University of Udine, Italy)
Business Science Reference • ©2018 • 190pp • H/C (ISBN: 9781522539872) • US $145.00

Handbook of Research on Applied Optimization Methodologies in Manufacturing Systems
Ömer Faruk Yılmaz (Istanbul Technical University, Turkey & Yalova University, Turkey) and Süleyman Tüfekçí (University of Florida, USA)
Business Science Reference • ©2018 • 449pp • H/C (ISBN: 9781522529446) • US $275.00

Grassroots Sustainability Innovations in Sports Management Emerging Research...
Marco Tortora (University of Florence, Italy)
Business Science Reference • ©2018 • 130pp • H/C (ISBN: 9781522535003) • US $145.00

Motivationally Intelligent Leadership Emerging Research and Opportunities
Michael A. Brown Sr. (Florida International University, USA)
Business Science Reference • ©2018 • 139pp • H/C (ISBN: 9781522537465) • US $155.00

Novel Six Sigma Approaches to Risk Assessment and Management
Vojo Bubevski (Independent Researcher, UK)
Business Science Reference • ©2018 • 251pp • H/C (ISBN: 9781522527039) • US $200.00

For an entire list of titles in this series, please visit:
https://www.igi-global.com/book-series/advances-logistics-operations-management-science/37170

IGI Global
DISSEMINATOR OF KNOWLEDGE

701 East Chocolate Avenue, Hershey, PA 17033, USA
Tel: 717-533-8845 x100 • Fax: 717-533-8661
E-Mail: cust@igi-global.com • www.igi-global.com

Table of Contents

Preface

In this second decade of the century, many manufacturing and service organizations globally are faced with demands for higher performance on returns, changes in the regulatory environment, new forms of competition, and difficult economic circumstances. A systematic approach through Lean Six Sigma is currently used by many organization, eliminating non-value added activities throughout their production system, as a solution to the scarcity of resources. Experience shows that both immediate and longer-term success with Lean Six Sigma requires proper grounding in concepts and off-line practices with the proper tools, methods, and applications to real-world problems. The book offers a review and analysis of the needed lean tools to improve the quality of inputs and outputs by identifying and analyzing methodologies that help to remove the causes of defects (errors) and, consequently minimize variability in business processes. The authors, Edem Tetteh, Ph.D. and Hans Chapman, Ph.D., devote the book to Lean Six Sigma concepts and practical set of quality management methods, including statistical applications, assessment procedure and special infrastructure of people within the organization. The authors believe that these resources can be used to define the sequence of systematic steps for practical solutions to complex organizational issues. Therefore, many known issues that will impede organization's quality triumph are tackled in the following chapters of the book.

Drs. Tetteh and Chapman investigate in Chapter 1 the rework performance in manufacturing using a lean process improvement. Many manufacturing facilities accumulate scrap with the hopes of reworking in order to still maintain a profit. The authors identify the factors in the process for performance rework that are causing the most cost to a local manufacturing company in an effort to reduce the cost and improve the process. The performance rework process is an in-depth process which includes inventory, assembly, test for noise, holding, and potential rework or scrap. In this chapter, the data comes directly from the facility as historical data from the year 2016. The authors also collected data on low and high band noise defects. Multiple methods were used such as value stream map, current state map, and future state map; and also the measurement of noise levels of the bearings using the Anderon meter and Waveometer. The third method consisted of constructing control charts using Minitab software. The authors found that bearings which fail due to low band noise, produce a low final yield after being reworked. On the other hand, bearings that fail due to high band noise had a much higher final yield after being reworked. Based on the finding, Drs. Tetteh and Chapman provide recommendations to manufacturing facilities on how to approach their low and high band scrap after Lean Six Sigma approaches were taken.

Drs. Tetteh and Chapman use lean tools to assess and improve lighting condition on a college campus in Chapter 2. They argue that results of the outdated lighting infrastructure, poor lighting conditions are evident in certain sections of the campus, given that some areas are not well lit, and can constitute unsafe environments for faculty, staff, students, and visitors. In addition to the poorly lit areas, the chapter shows that the current lighting near older campus buildings is not energy efficient. Consequently, the lack of energy efficiency could potentially be imposing additional costs to the university when looking at the extent of the electricity usage. The authors use several lean tools to collect data on the light luminosity, pole locations using Global Positioning System (GPS) coordinates, and other conditions. The assessment and analysis of the data help to propose lean based recommendations that require the implementation of new and additional lighting for better efficiency and safety.

Lean principles were used by the authors to ensure the quality level in the design of a virtual learning tools. Chapter 3 clarifies the relations between Lean and Six Sigma as components of a quality system showing that both concepts are not interchangeable. Lean focuses on the elimination of waste, non-value added tasks, and variation, with the overall goal of continuous improvement. On the other hand, Six Sigma focal point is the depiction of commitment to lean. Six Sigma tends to be data driven and aims at addressing complex business problems through the use of the five-phase method approach. Lean Six Sigma can be applied to any form of business to facilitate the optimization, and streamlining of the internal processes. Drs. Tetteh and

Chapman employ lean principles in the design of the virtual tool to increase efficiency, improve quality, and provide better user experience to potential users. The learning tool allows the virtual exploration of a warehouse through the use of an avatar. Basic warehousing operations are designed using a Lean Six Sigma principle such as the Plan-Do-Check-Act in a way to enhance the teaching and learning experience of the users.

The topic of reduction of hospital costs through a better management of supply waste is discussed in Chapter 4. Hospitals lose a large amount of money through waste as defined by lean. Processes have been in place many years at medical facilities and waste is often accepted as unavoidable in the rightful "patient first" mentality. The authors argue that by adapting and utilizing simple lean and statistical tools such as scatter diagrams, histograms, Pareto charts, fishbone diagrams and control charts, waste can be identified and reduced or eliminated. This leads to increased satisfaction, efficiency and creation of a competitive advantage.

Lean and Total Quality Management (TQM) in civil engineering is introduced in Chapter 5. Drs. Tetteh and Chapman examine the utilization of TQM, ISO 9000 and Lean concepts in the management processes of civil engineering consulting firms. The chapter briefly discusses the history of management theory and the existing literature related to Total Quality Management, ISO 9000 and Lean concepts as applied to the civil engineering profession and investigates the practical application of these concepts in civil engineering firms' operations, project management and the extent that academic courses in these topics are offered in colleges and University. The authors use internet research and interviews with civil engineering managers and college professors to uncover that while the literature strongly endorses the utilization of these management concepts, the actual use and recognition is little known in the professional business of civil engineering. Also, in general, undergraduate civil engineering academic programs in the United States do not offer specific courses, which cover these concepts in their curriculums.

Chapter 6 takes readers on a journey to the heart of the continuous improvement process in a lean enterprise. Drs. Tetteh and Chapman highlight the fundamental role played by lean principles in the promotion of continuous improvement in the workplace. In that regard, the chapter assesses the impact of Kaizen principles in an organization's strategy to increase productivity and effectiveness in its operations. The authors use a Single Minute Exchange of Dies methodology to dramatically reduce the time it takes to complete equipment changeovers in a world-class manufacturer plant located in the United States. A Single-Minute Exchange of Dies system combined with other lean tools, such as 5S and Standard Operating Procedures (SOP) were used to convert as many changeover steps as possible to tasks that can be performed while the equipment is running. Furthermore, the remaining vital steps of the process were simplified and streamlined. Drs. Tetteh and Chapman support the criticality of the concept of Kaizen coupled with a Single-Minute Exchange of Dies system as a very effective method to achieve continuous improvements.

Chapter 7 introduces Dell's Just-In-Time system and how the model can be implemented in the country of Ghana in West Africa. Dell Computer Corporation founded by Michael Dell started in 1985 and become the world's leading computer systems company with a simple motto to design build and customize products and services to satisfy a range of customer requirements. The authors show that a few years after introducing its next-day on-site service, which will become the company core of its competitive edge, Dell started to expand internationally with the opening of many subsidiaries. In this chapter, Drs. Tetteh and Chapman assess the feasibility of the implementing of Dell's model in Ghana and the evaluation of the impacts this implementation will have on an emerging economy such as Ghana through an assessment of key factors such as Ghana's rich natural and human resources.

Overall, the seven chapters provide strength and quality to this book to showcase strategies to improve process efficiency and effectiveness in business processes using Lean Six Sigma principles and applications. The book emphasizes the potential use of Lean Six Sigma to improve business, human and machine system resources management with the overarching objective of avoiding delays, reducing errors, speeding up the time to implementation, and improve operations and changes. Essentially, the book presents the detailed description of the main principles, practices, and methodologies used in Lean Six Sigma and Quality Management Systems as applies to organizational decision making. Process-oriented nature of quality as applies to complex business processes is highly emphasized.

Organizational professionals and researchers working in the fields of process and business management, both from a theoretical and practical point of views, are the main target audience. Furthermore, the book provides insights to business executives responsible for understanding tools and technology to positively impacting every aspect of the business. Faculty in education, business, engineering, and technology will use this book as a good resource for case studies. Nonetheless, other than the groups listed above, the general population and organizations involved in business management locally or globally including researchers are also the targets of the book.

Edem G. Tetteh
Rowan College at Burlington County, USA

Hans Chapman
Cape Fear Community College, USA

Acknowledgment

We would like to express our profound gratitude to the many people who saw us through the preparation of this book, all those who provided support, talked things over, read, wrote, offered comments, allowed us to quote their remarks, and assisted in the editing, proofreading, and design. Among them are colleagues and students who assisted with the studies for the chapters such as Dr. Coray Davis and Dr. Keith Williamson, Mr. Marvin D. Mills, Mr. Anthony Minggia, Mr. Brian C. Elam, Ms. Deshaney Bigelow, and Mr. Mario Frater from Virginia State University. Special credit goes to recent Morehead State University graduates: Paula Sloas whose work contributed to the chapter on Lean Process Improvement for Rework Performance in Manufacturing, Ricky Hamilton whose work contributed to the studies on Reducing Hospital Costs by Reducing Supply Waste, and Mr. Joe Curd of Morehead State University for his work involving the chapter on Lean and Total Quality Management in Civil Engineering. Also, the book development team at IGI Global, for helping us to stay on track and providing much needed guidance and feedback.

Above all, Dr. Tetteh wants to thank his wife, Momo, and their children, Shania, Kyle, and Evan, who supported and encouraged him in spite of all the time that was spent away from home during the preparation of this book. It was a long and difficult journey for them. Dr. Chapman would like to thank his wife, Ophelia, and their children Harold, Hathlyn, and Hannah for the love, patience, understanding, and faithful support shown while preparing this book. Finally, we ask for forgiveness for forgiveness from all those who provided useful insights over the course of the book preparation and whose names we have failed to mention. We would like to express our heartfelt gratitude for your efforts and assistance during the book project.

Edem G. Tetteh
Rowan College at Burlington County, USA

Hans Chapman
Cape Fear Community College, USA

Chapter 1
Lean Process Improvement for Rework Performance in Manufacturing

ABSTRACT

The primary purpose of this chapter is to identify the factors in the process for performance rework that are causing the most cost to a local manufacturing company to reduce the cost and improve the process. The data comes directly from the facility as historical data from the year 2016. The authors also collected data on low band and high band noise defects. The first method is the value stream map, current state map, and future state map. The second method is to measure noise levels of the bearings using the Anderon meter and Waveometer. The third method is to construct control charts using Minitab software. The findings demonstrated that bearings that fail due to low band noise produce a low final yield after being reworked. On the other hand, bearings that fail due to high band noise had a much higher final yield after being reworked.

BACKGROUND OF LEAN MANUFACTURING

The goal of lean manufacturing is to be highly responsive to customer demand by reducing waste. Lean manufacturing aims at producing products and services at the lowest cost and as fast as required by the customer. The lean concept originated in Japan after the second world war when Japanese

DOI: 10.4018/978-1-5225-4062-5.ch001

manufacturers realized that they could not afford the massive investment required to rebuild devastated facilities. Toyota is often credited for the creation of lean manufacturing. Lean manufacturing gives manufacturers a competitive edge by reducing cost and improving productivity and quality. It can include improvement in processing time, lead time, cycle time, set up time, inventory, defects and scrap, and overall equipment effectiveness.

Background of Quality Control

Quality control is the process that ensures customers receive products free from defects and that meet their needs. Some common tools used to support quality control include statistical process control (SPC) and Six Sigma. Statistical process control monitors and controls quality by tracking production metrics. It helps quality managers identify and solve problems before products leave the facility. Six Sigma is a systematic approach for eliminating errors. It uses statistical methods to improve quality by minimizing variability.

Purpose of the Research

The primary purpose of this research is to identify the factors in the process for performance rework that are causing the most cost to the manufacturing company. The goal is to use these factors to reduce the cost and improve the process. The performance rework process is an in-depth process which includes inventory, assembly, test for noise, holding, and potential rework or scrap.

Objectives of the Research

- To appraise the existing Performance Rework Process at a local manufacturing company.
- To accurately measure noise levels of parts or parts processed, such as standard cylindrical series, RB (rubber mounted inserts), ER (non-spherical outer diameters), and SK (special order) Inners/Outers for performance ball bearings.
- To measure and analyze output of noise levels in hz, for variability using control charts in Minitab statistical software.
- To improve and control the Performance Rework Process.

- To develop and standardize sorting process to meet customer demand.
- To develop a more cost-efficient process to meet customer demand.

Significance of Research

The significance of looking into the performance rework process is that many manufacturing facilities have the issue of scrap and rework. This is a topic that can resonate across many different companies. By establishing the areas that can improve the process, companies can then adjust their processes in order to decrease costs and improve customer satisfaction. The results of this research would be of value to organizations who want to find out if it is worth the cost to rework or to send everything nonconforming to scrap.

Assumptions and Limitations

The main limitation of this study is lack of historical data. While the factory has historical data on scrap, there is no current data for rework fallout. This means that parts could still be non-conforming even after being reworked. Also, the data collected by the author used small sample sizes of 4 and a small lot size of 34 for one of the two scrap reasons studied. The limitation of small sample sizes and small lot sizes should be observed.

Definition of Terms

- Lean Manufacturing aims at producing products and services at the lowest cost and as fast as required by the customer. The goal of lean manufacturing is to be highly responsive to customer demand by reducing waste (Bhamu et al., 2014).
- Quality Management consists of activities and functions involved in determination of quality policy and its implementation through means such as quality planning and quality assurance (Quality Management, 2017).
- Total Quality Management is an effort in which all members of an organization participate in improving processes, products, services, and the culture in which they work (Total Quality Management, 2017).
- Kaizen is continuous, incremental improvement of an activity to create more value with less waste (Womack, 2003).

- Kanban is a small card attached to boxes of parts that regulates pull in the Toyota Production System by signaling upstream production and delivery (Womack, 2003).
- Six Sigma is a disciplined, data-driven approach and methodology for eliminating defects (driving toward six standard deviations between the mean and the nearest specification limit) in any process (What is Six Sigma? 2017).
- Lean Six Sigma is a managerial approach that combines Six Sigma methods and tools and the lean manufacturing/lean enterprise philosophy, striving to eliminate waste of physical resources, time, effort and talent, while assuring quality in production and organizational processes (Lean Six Sigma Definition, 2014).
- Attribute Charts plot characteristics for data in categories such as defective, non-defective, conforming, and non-conforming. These charts measure defects in counts, units, and proportions (What are Attributes Control Charts? 2017).
- Statistical Process Control is the use of valid analytical statistical methods to identify the existence of special causes of variation in a process (Quality American, Inc., 2013).
- Takt Time is the available production time divided by the rate of customer demand. This sets the pace of production to match the rate of customer demand and becomes the heartbeat of any lean system (Womack, 2003).
- Value Stream Mapping is the identification of all the specific activities required to design, order, and provide a specific product, from concept to launch, order to delivery, and raw materials into the hands of the customer and displayed in the form of a map. (Womack, 2003).
- Rework in Manufacturing is the process of rectifying mistakes made during manufacture so that a product meets requirements. It could be as simple as affixing a new label, or as extensive as re-machining (Socius, 2017).
- DMAIC is an acronym for Define, Measure, Analyze, Improve and Control which refers to a data-driven quality strategy used to improve processes. It is an integral part of Six Sigma methodology (The DMAIC Process, 2017).

LITERATURE REVIEW

Lean Philosophy

Lean management is an applied philosophy that many manufacturing organizations have adopted to obtain the flexibility needed to meet new competitive challenges, such as, eliminating waste, enhancing production speed, and pushing innovation (Demers, 2002). The objective of the Lean approach is to respond rapidly to changing client tastes and to offer the most value possible at mass production costs. A number of management approaches have similar objectives. Total Quality, Kaizen, Kanban, and Six Sigma are all solutions that work towards an equivalent goal.

After World War II, the Japanese adopted Henry Ford's mass production principles. Taichii Ohno made a major contribution to these changes when it developed the Toyota Production System. This system is the production model that everyone seeks to emulate even to this day. The system attempts to achieve three goals: flow, harmony (or Takt Time) and synchronization (pull flow). These are some of the main principles behind the Lean approach (Demers 2002).

Womack (2003), outlined five basic principles associated with Lean. A company must: specify value from the standpoint of the customer; identify the value stream for each product family; make the product flow; allow the customer to pull production forward; and work toward perfection. Notice that in this process, the customer is the focal point right from the beginning. If a company can follow this method, they will be successful from a Lean standpoint.

Lean Six Sigma Principle

Lean Six Sigma is one of the most used hybrid methodologies for continuous improvement of manufacturing processes. Bhuiyan and Baghel (2005) describe Lean Six Sigma as a combination of Lean Production and Six Sigma. The advantage with the combination of two different concepts is that it can facilitate solving problems that would be difficult to solve if the concepts were used separately.

Using Lean Six Sigma makes it possible to manage and solve different problems since the user can choose among different tools (Salah, Rahim and Carretero, 2010). The core of Lean Production is to reduce waste, and the method and tools of bringing the process into statistical control originates from Six Sigma. Using these methods and tools together makes it possible to achieve both cost reduction and quality improvements. Figure 1 represents tools typical for Lean Production, Six Sigma, and Lean Six Sigma.

Lean Tools

Breyfogle (2003) describes Value Stream Mapping as a tool that is used to discover where effort should be directed in order to improve a business. VSM is a well-recognized tool in Lean Production. The tool helps the company to map and later improve a production process regarding both material and information flow (Chen, Li and Shady, 2010). See Figure 2 to review the symbols used in a VSM. Also, see Appendix B for a more in depth review.

Figure 1. Tools that are typical for Six Sigma, Lean Production, and Lean Six Sigma (Salah et al., 2010)

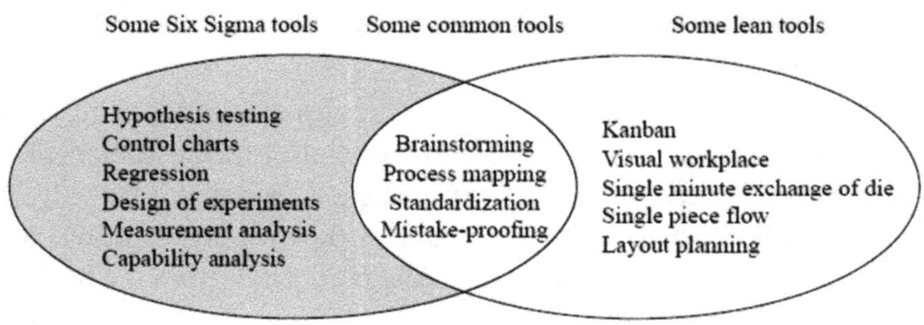

Figure 2. Symbols for VSM-configuration (Bellgran and Safsten, 2005)

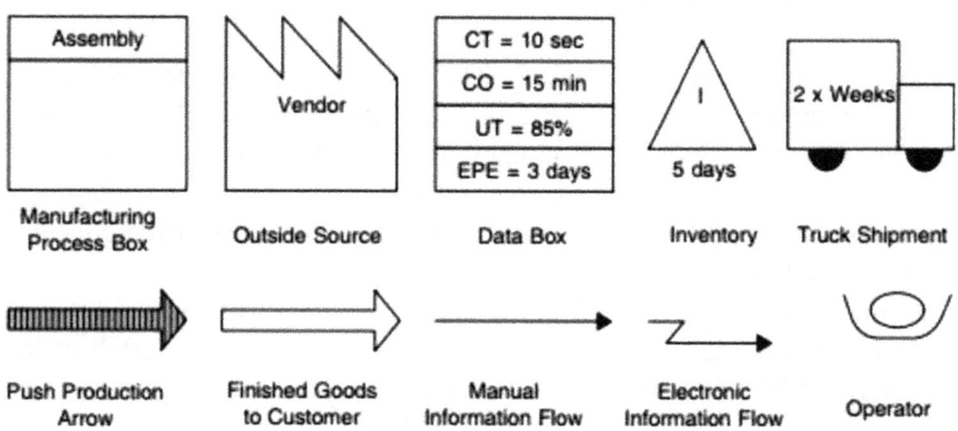

A VSM must first map the current state before it can map the future state. The Current State Map (CSM) is based on collected data and information of the current state. The information required to establish a process map can be gathered by observing the process while in progress. The observation can be conducted by a walkthrough. Depending on what information is required, the walkthrough can be performed downstream or upstream of the material/information flow.

The last thing included in the process map is the timeline below the activities, which contains the lead time of each activity and how many operators they require (Bellgran and Safsten, 2005). The lead times can be value-adding, non-value-adding, or necessary non-value-adding. When the lead times are established and categorized, their difference can be calculated. The calculation describes the amount of value-adding activities compared to the amount of non-value-adding activities in the whole process.

Rework Process

Manufacturing operations are often imperfect and thus their outputs may contain defective items. These defective items can either be scrapped or reworked. In some cases, they can be sold at a reduced price. However, for the factory that is being used for this research, selling defective items at a reduced price is not an option. In all cases, substantial costs result. In the case where the items can be reworked, the costs consist of the additional resources needed to rework the items, and the costs related to the longer turnaround time resulting from rework. The yield of a manufacturing process is often a function of the lot size used. Therefore, varying the lot size can have an impact on the productivity of the manufacturing operation.

Six Sigma

Aside from being a measure of variability and organization's quality performance, Brue and Howes (2006) mention that Six Sigma is also a management philosophy. It is a strategy as well as a problem-solving and improvement methodology which can be applied to every type of process to eliminate the root cause of defects. Some of the main benefits that an organization can gain from applying Six Sigma are cycle time improvements, cost reduction, defect elimination, an increase in customer satisfaction, and a significant rise in profits.

In an article by Kumar et al (2015), they demonstrated the empirical application of Six Sigma and DMAIC to reduce product scrap within a piston manufacturing organization. Through their research and analysis, they discovered that the design of casting spoon and its material influenced the amount of defective pistons produced. As a result, they reduced their scrap percentage from 9.9% to 5%.

In a case study by Valles et al (2009), a Six Sigma project was conducted at a semiconductor company dedicated to the manufacture of circuit cartridges for inkjet printers. This company was having trouble with electrical failures and this problem accounted for about 50% of all defects. The case study's main question was, "what is causing this problem?" Since this issue was the root cause of 50% of their defective parts, it was crucial to locate the main problems, causes, and actions to reduce the level of defects.

They were able to find that abrasive pressure, height of the tool, and cycle time were the main factors that caused the defects. By improving these, the results were a reduction in the electrical failures of around 50%. Thus, resulting in happier customers and less waste for the company.

Six Sigma DMAIC

One of the distinctive approaches of Six Sigma is DMAIC. The DMAIC model refers to the five interconnected stages (define, measure, analyze, improve, and control) that systematically help organizations to solve problems and improve their processes.

The Define stage within the DMAIC process involves identifying, evaluating, and selecting projects for improvement as well as selecting teams. The Measure stage includes collecting the data on size of the selected problem, identifying key customer requirements, and determining key project and process characteristics. The Analyze stage focuses on analyzing the data, and establishing and confirming the most important determinants of the performance. The Improve phase focuses on designing and carrying out experiments to establish cause and effect relationships as well as optimizing the process. Finally, the Control phase involves designing the controls, making improvements, and implementing and monitoring those improvements.

When improving a current process, if the problem is complex or the risks are high, DMAIC is an excellent go-to method. If the risks are low and there is an obvious solution, some of the DMAIC steps could be skipped. However, this is not advised. There are two main approaches to implementing DMAIC.

Figure 3. DMAIC cycle (Brue & Howes, 2006)

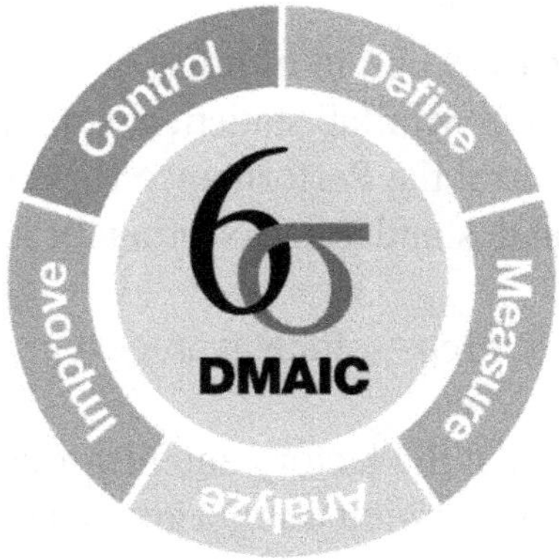

The first is the team approach in which individuals who are skilled in the tools and method, such as quality or process improvement experts, lead a team. The team members work on the project part-time while caring for their everyday responsibilities. The quality or process improvement expert might be assigned to several projects (Berardinelli, 2012).

The second tactic involves the kaizen event method, an intense progression through the DMAIC process typically done in about a week. Prep work is completed by the quality or process improvement expert, and is centered on the define and measure stages. The rest of the stages are done by a team of individuals who have been pulled from their regular duties for the duration of the kaizen event.

METHODOLOGY

Multiple methods have been selected to measure and analyze the performance rework process at a manufacturing facility. The primary goal of this research is to measure and analyze the current process, discover what scrap reason is costing the company the most money, and make suggestions for improvement. The research objectives are as follows:

- To appraise the existing Performance Rework Process at a local manufacturing company.
- To accurately measure noise levels of parts or parts processed, such as standard cylindrical series, RB (rubber mounted inserts), ER (non-

spherical outer diameters), and SK (special order) Inners/Outers for performance ball bearings.

- To measure and analyze output of noise levels in Hertz, for variability using control charts in Minitab statistical software.
- To improve and control the Performance Rework Process.
    - To develop and standardize sorting process to meet customer demand.
    - To develop a more cost-efficient process to meet customer demand.

The data was collected at the factory and historical data was provided by the factory. Appropriate permission was not granted for use of any identifiable factors for the facility studied within this research. Therefore, both the name of the state and the specific manufacturing facility will be omitted. The data will provide: the number of scrap pieces, cost of scrap pieces, reason codes of scrap pieces. All of this data is historical from the year 2016. Data collection included low band and high band scrap.

The nature of the data collected will be attribute and binomial since the data was not collected over time. Attribute data in the form of pass or fail will be used. Data will be organized using Microsoft Excel in order to further measure and analyze the data utilizing the Minitab 17 software program. The data will be analyzed using pareto charts, histograms, a pie chart, and P-charts. Pareto charts are primarily used as defect identifying and tracking tools.

Method One: VSM

The first objective involves the appraisal of the existing process for performance rework. In order to do so, a Value Stream Map (VSM) - current state process map was created. A flowchart that displays the existing process was also created. The current state process map can be used to determine a plan of action to address opportunities for immediate improvement in order to get to the ideal state for the internal reworking process. Figure 6 displays the current state process map for the hone rework process.

A Kaizen event was led to determine the Future State of VSM. The immediate improvements included organization of the parts on the Hone rack. Figure 4 shows a before kaizen photo of the hone rack. After the kaizen and the results of this study, the rack was eliminated and a cart is now being used (Figure 5). A 5S layered process audit was written and conducted to improve organization at the hone rack and thus maintain the Future State VSM once achieved.

Figure 4. Rack holding the parts waiting to be reworked on hones. Photo taken by author at facility where research was conducted

Figure 5. Cart now used for parts waiting to be reworked on hones. Rack was eliminated during Kaizen. Photo taken by author at facility where research was conducted.

Figure 6. Hone Rework Process Current State Map. VSM created in Microsoft Office Visio.

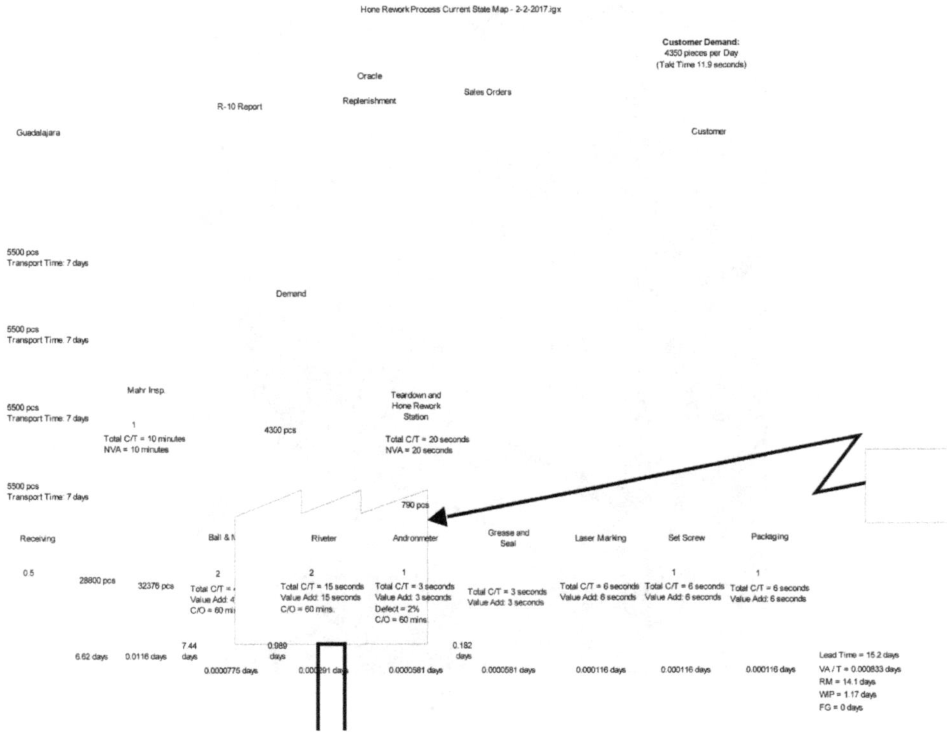

Method Two: Anderon Meter and Waveometer

The second objective is to measure noise levels of the bearings. It is impossible to see the bearings inside once they are assembled. The noise levels of bearings must be checked when they are rotating. An Anderon meter is used to check the noise of complete bearings. It tells the noise level of each bearing with a certain unit. The Anderon meter is used on the shop floor. When noise is checked by the Anderon meter, a specified preload is applied to each different bearing. This preload is determined by the size and type of bearing. "Mechanical vibration of the bearing outer ring is detected in the radial direction by contacting the outer ring with a velocity pickup, and the signal from the pickup is converted to a variable electric output" (Ball Bearing Vibration and Noise, 2016). The inner ring of the bearing is rotated at 1800 RPM and the outer ring remains stationary during the measurement. Vibrations from other sources can be eliminated in this method.

Figure 7. Hone Rework Process Current Flow Chart. Image created in Microsoft Office Visio.

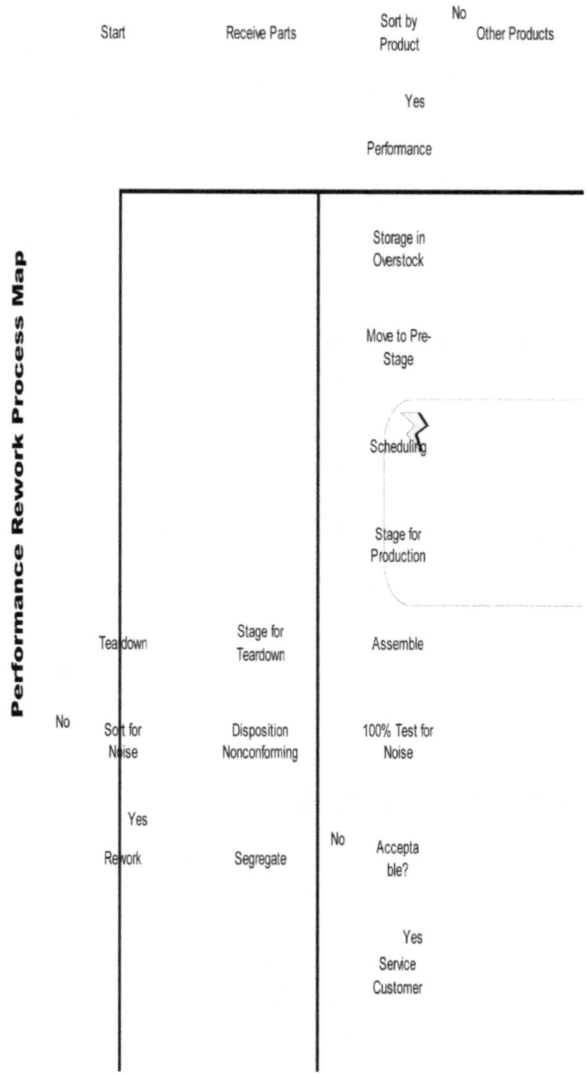

The vibration noise value is shown on the meter after all frequencies are separated into three bands. Low band frequency is 50 – 300 Hz. Medium band frequency is 300-1800 Hz. High band is 1800-10000 Hz. Figure 9 shows a close up of an Anderon meter. The only difference between this machine and the machine located at the factory, is that this machine has a mechanical pusher. At the facility where this study was completed, the operator physically pushes the bearing onto the Anderon meter.

Figure 8. Future State Hone Rework Process Flow Chart. Image created in Microsoft Office Visio.

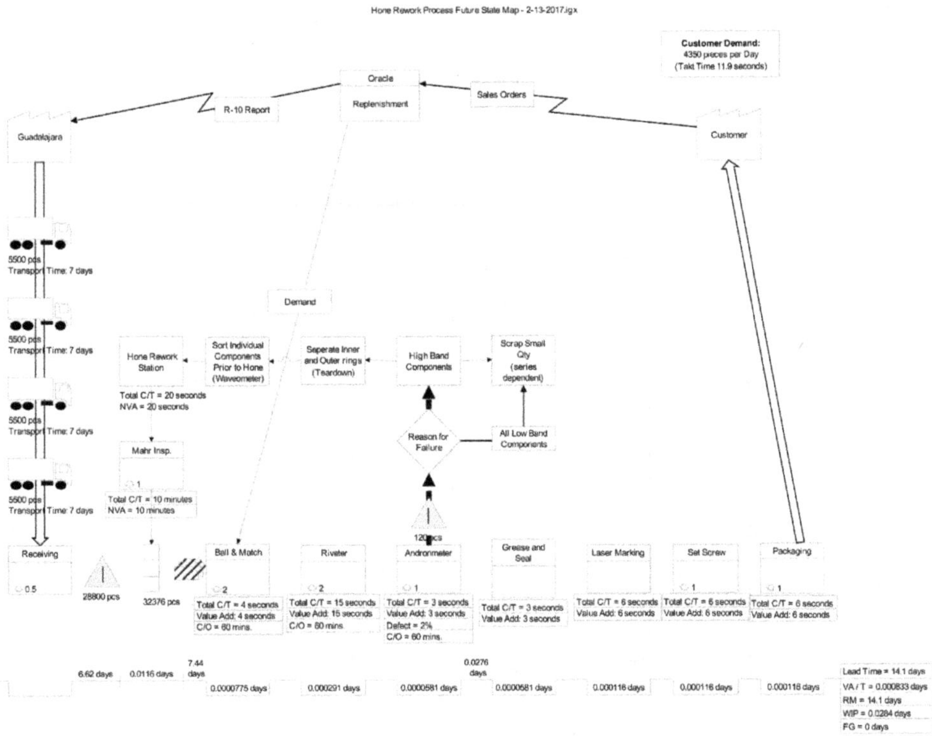

Figure 9. Anderon meter (Ball Bearing Vibration and Noise, 2016)

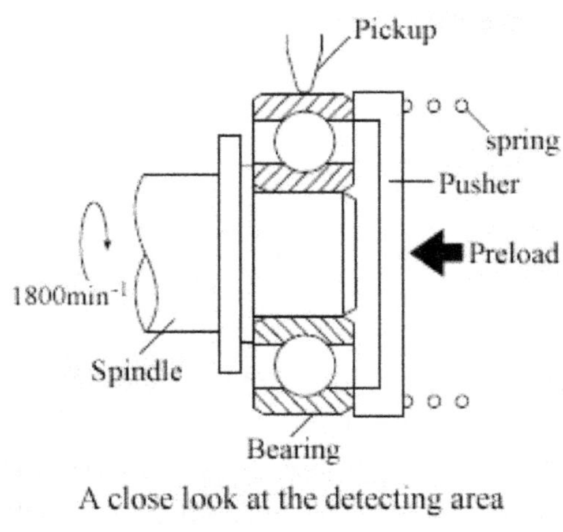

A Waveometer is a common measuring instrument. It senses and reports the amplitude of waves or ripples in the balls surface in frequency bands. This is very important in the rolling element of bearing applications because its amplitude has a direct effect on the vibration or noise level of a running bearing. The Waveometer is a production test machine which requires a special setup for each size ball.

Method Three: Minitab Control Charts

The third objective involves measuring and analyzing the noise outputs statistically. In order to do this, Minitab software will be used. Control charts, such as P charts, will be produced in Minitab in order to show whether or not the process being used for rework is controlled. A statistical analysis of the current performance rework process will show whether or not the process is capable and in control.

Method Four: Recommendations

Finally, the fourth objective is to improve and control the Performance Rework Process at a local manufacturing facility. Recommendations will be made based on the findings from the previous three methods. Development and standardization of the sorting process in order to meet customer demand will be examined. Also, developing a more cost efficient process in order to meet customer demand will be included in this stage. Following these developments, recommendations will be made to the company on any ways to improve and better control the process for performance rework.

DATA AND FINDINGS

Data Collection

All of the data was collected at the factory where this research was conducted. Two sets of data were collected. The main two types of defects that were studied in this research were high band and low band. High band and low band are noise levels that are measured on the Anderon meter. The factory noticed that they had several parts on a rack waiting to be reworked on the hone machines. The majority of these parts had the scrap reason code of low band or high band. Therefore, these two reasons were studied to determine if the parts on the rack could be reworked on the hones and still produce a viable end result.

Figure 10 displays the flowchart of the hone rework process. This begins from receiving the parts, then they are sorted by product. This research looked only at performance ball bearings. They are pre-staged and then sent to the staging area which is temperature controlled. The line operator looks at the manufacturing orders to see what parts are needed. Then, the parts are staged for assembly and run through the assembly line. They are checked on the Anderon meter and if they are acceptable, they service the customer. If they are rejected, then they are placed in a bin and sent to the hold area. After they are approved by the supervisor and signed off on, they are sent to teardown. They are torn down and checked on the Waveometer. Then they were originally placed on the hone rack. They awaited there to be reworked on the hone.

Figure 11 shows the harmonics of a low band bearing. As shown, the low band can be in the form of three lobes. Figure 12 shows what high band looks like. The image is actually low band and high band. The high band is shown in the form of the rigid lines that resemble a pop cap. The research specifically looked into these two defects to see if they could be reworked or not.

An Excel spreadsheet was used to store the collected data for further analysis. Data was collected on low band and high band. The facility has a hold area where parts wait before being torn down and reworked on the hones. From this hold area, 106 bearings with the reject code of low band were taken to be reworked. 106 inners and 106 outers (212 total pieces) were honed. This took the hone operator approximately 11 hours to complete. After these pieces were reworked they were reassembled and checked on the Waveometer. Of the 106 inners that originally failed for low band noise, after being reworked on the hone 40 inners passed on the Waveometer. This is a 37.7% pass rate. Of the 106 outers that originally failed for low band noise, after being reworked on the hone 13 outers passed on the Waveometer. This is a 12.2% pass rate. The 13 good outers were paired with 13 of the good inners and of those 13 complete bearings, only 3 produced a bearing with a usable internal clearance match. This is a 2.8% Final Yield.

From the hold area, 34 bearings with the reject code of high band were taken to the hones to be reworked. 34 inners and 34 outers (68 total pieces) were honed. After these pieces were reworked, they were reassembled and checked on the Anderon meter. However, too much material had been taken off of the parts during the hone process and due to this there were not balls of

Figure 10. Flowchart with photos of hone rework process. Photos and chart created by author

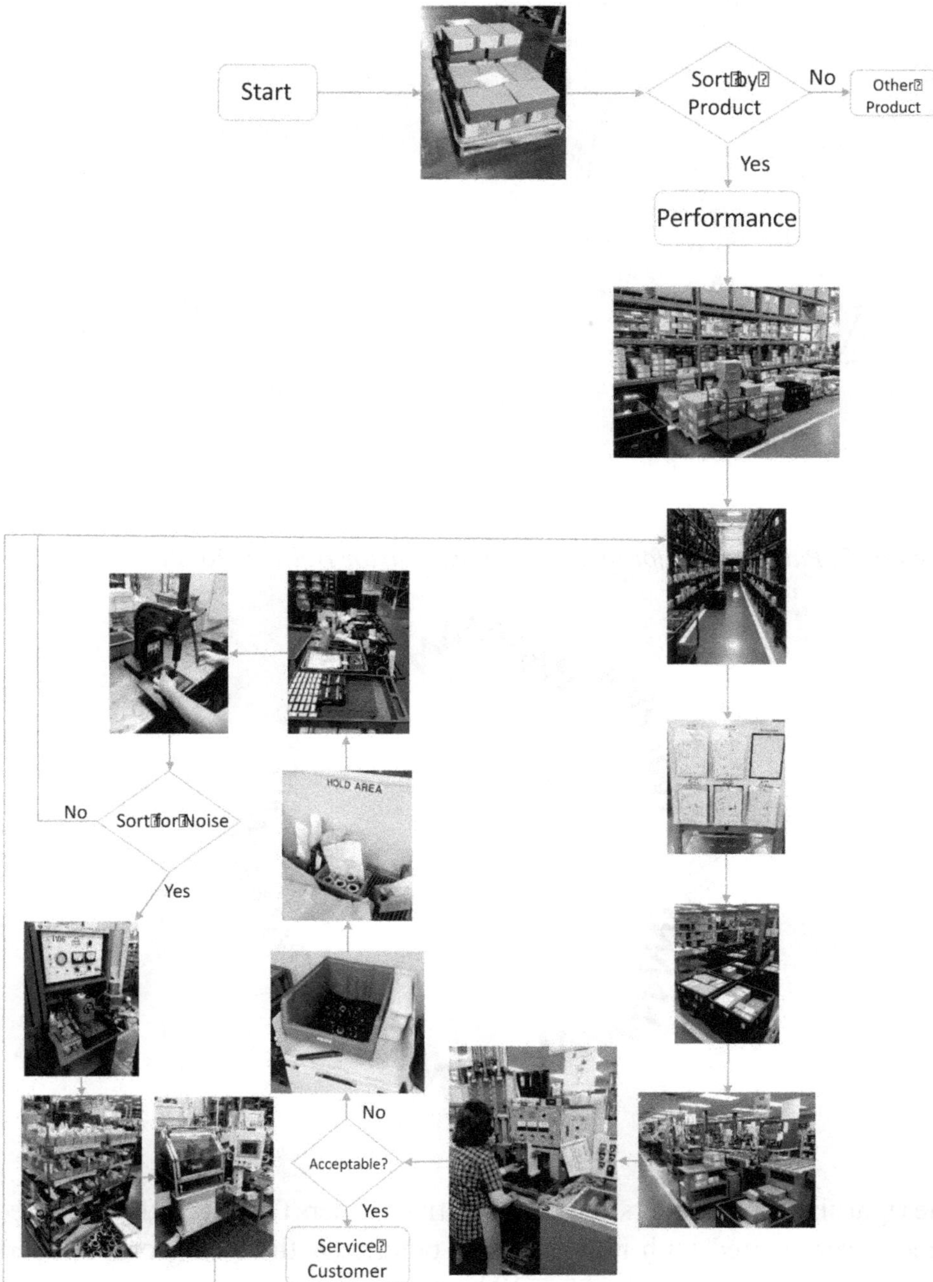

Figure 11. Part rejected for low band noise (Taylor Hobson, 2017)

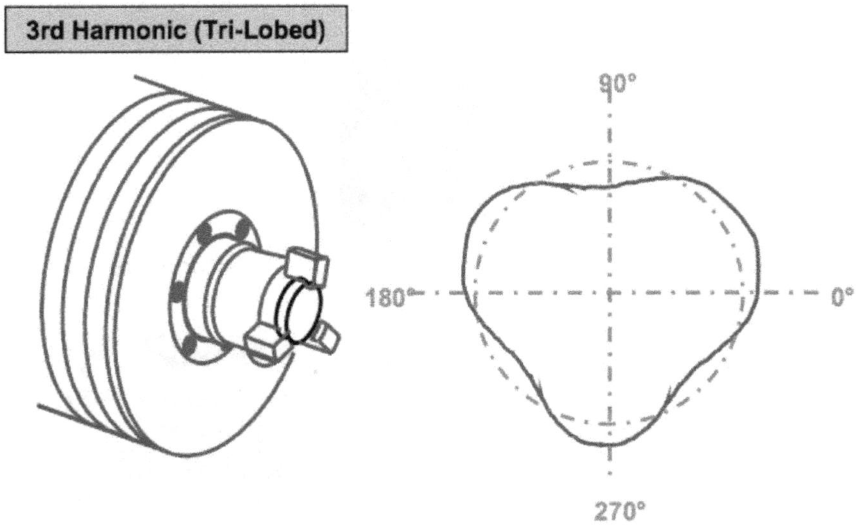

Figure 12. Part rejected for high band noise (Taylor Hobson, 2017)

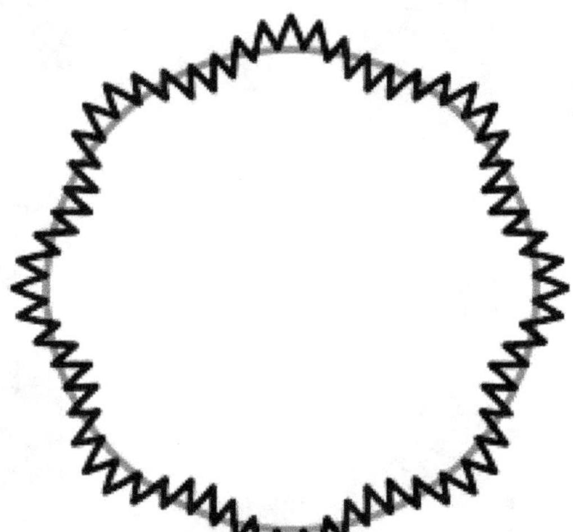

the right size for the inners and outers to fit together. Therefore, the reworked inners were paired with non-reworked outers. Of the 34 reworked inners paired with non-reworked outers, 19 bearings passed on the Anderon meter. Of the 34 reworked outers with non-reworked inners, 34 bearings passed on the Anderon meter.

The data was entered into an Excel spreadsheet. A sample size of 4 was used for the low band data. After the data was grouped, it was entered into Minitab 17 software to generate p charts. The p chart of low band reworked inners (Figure 13) shows that there were two points that were out of the control limits. As you can see from Table 1, point 7 and point 20 were out of control as 100% of the parts in those subgroups were defective. The *p* value for each subgroup was found by using the formula, $p = \frac{np}{n}$ where np is the number of defectives and n is the total number of parts sampled. The p-bar for the low band reworked inners is 0.611 meaning that 61% of these inners were nonconforming even after being reworked.

The p chart of low band reworked outers (Figure 14) shows that all points are within the control limits. The p-bar value for the low band reworked outers was 0.491, meaning that 49% of the outers were nonconforming after being reworked.

The 13 outers that were acceptable were paired with 13 of the inners that passed on the Waveometer to produce 13 bearings. They were categorized by a subgroup of 4 and the p chart for the bearings is below (Figure 15). The p-bar value is 0.8125, indicating that 81% of the final bearings were defective.

Table 1. Subgroups and P Values of Low Band Reworked Inners

Group	*P*	Group	*P*
1	0.50	15	0.25
2	0.75	16	0.75
3	0.75	17	0.50
4	0.75	18	0.75
5	0.75	19	0.50
6	0.50	20	1.00
7	1.00	21	0.75
8	0.50	22	0.50
9	0.50	23	0.50
10	0.75	24	0.25
11	0.50	25	0.75
12	0.75	26	0.75
13	0.75	27	0.00
14	0.50		

Figure 13. P Chart of Low Band Reworked Inners. Chart constructed in Minitab 17 software program by author. Data collected by author from manufacturing facility. Retrieved March 7, 2017.

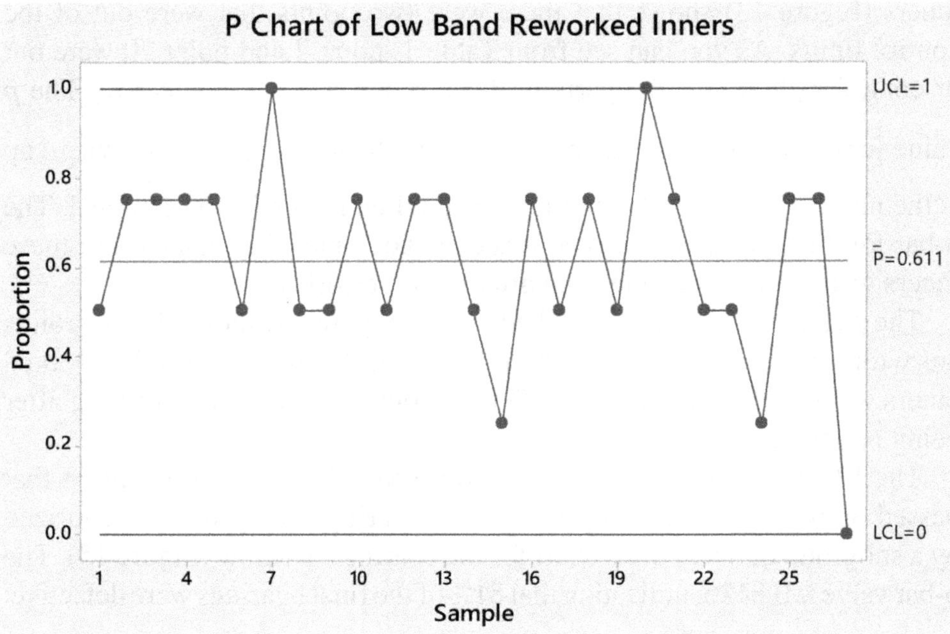

Table 2. Subgroups and P Values of Low Band Reworked Outers

Group	*P*	Group	*P*
1	0.75	15	1.00
2	0.75	16	0.75
3	1.00	17	1.00
4	0.75	18	1.00
5	1.00	19	1.00
6	1.00	20	0.75
7	0.75	21	1.00
8	1.00	22	0.75
9	0.75	23	1.00
10	0.75	24	0.75
11	1.00	25	1.00
12	0.75	26	0.75
13	0.75	27	1.00
14	1.00		

Figure 14. P Chart of Low Band Reworked Outers. Chart constructed in Minitab 17 software program by author. Data collected by author from manufacturing facility. Retrieved March 7, 2017.

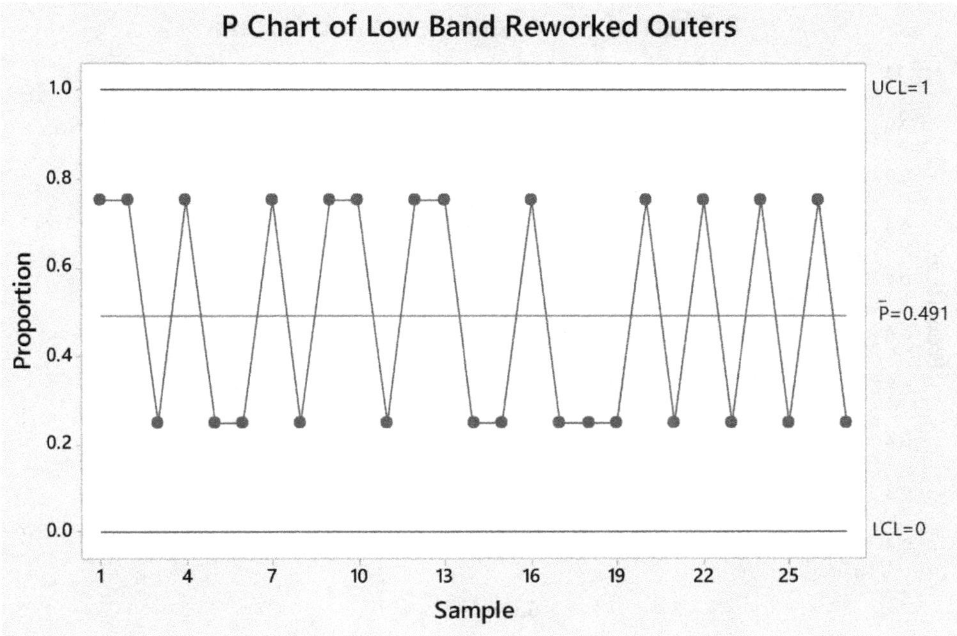

Table 3. Subgroups and P Values of Low Band Reworked Inners and Outers

Group	*P*
1	0.75
2	0.50
3	1.00
4	1.00

A sample size of 4 was used for the high band data, as well. After the data was grouped, it was entered into Minitab 17 software to generate a p chart. The p chart of high band reworked inners with non-reworked outers (Figure 16) shows that there were two points that were no points that were out of the control limits. The p-bar for the high band reworked inners with non-reworked outers is 0.417 meaning that approximately 42% of these inners were nonconforming even after being reworked and paired with non-reworked outers.

Figure 15. P Chart of Low Band Reworked Inners and Outers. Chart constructed in Minitab 17 software program by author. Data collected by author from manufacturing facility. Retrieved March 7, 2017.

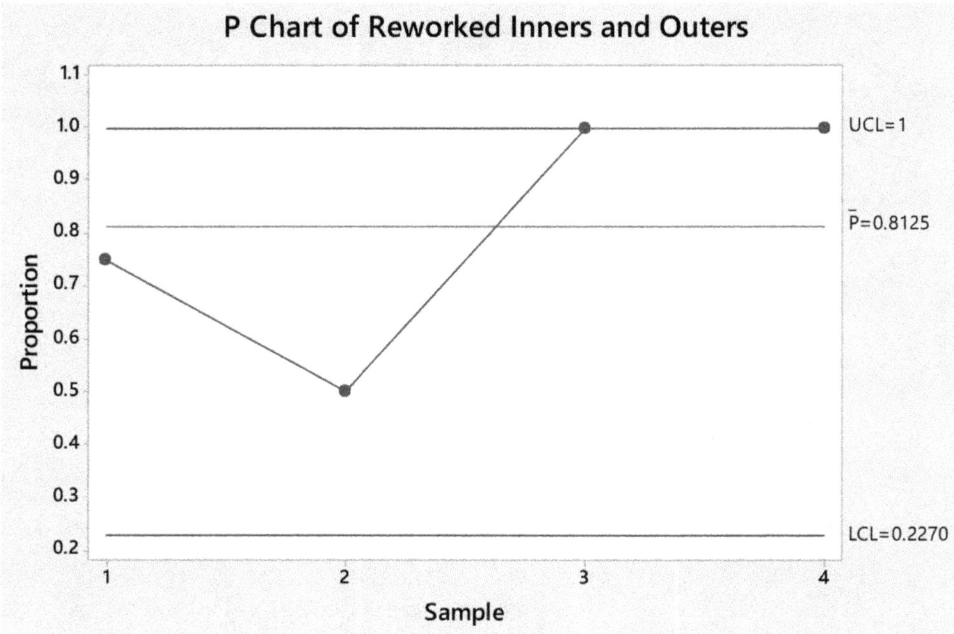

Table 4. Subgroups and P Values of High Band Reworked Inners and Non-Reworked Outers

Group	*P*
1	0.50
2	0.25
3	0.50
4	0.50
5	0.50
6	0.25
7	0.50
8	0.75
9	0.00

Figure 16. P Chart of High Band Reworked Inners with Non-Reworked Outers. Chart constructed in Minitab 17 software program by author. Data collected by author from manufacturing facility. Retrieved March 7, 2017.

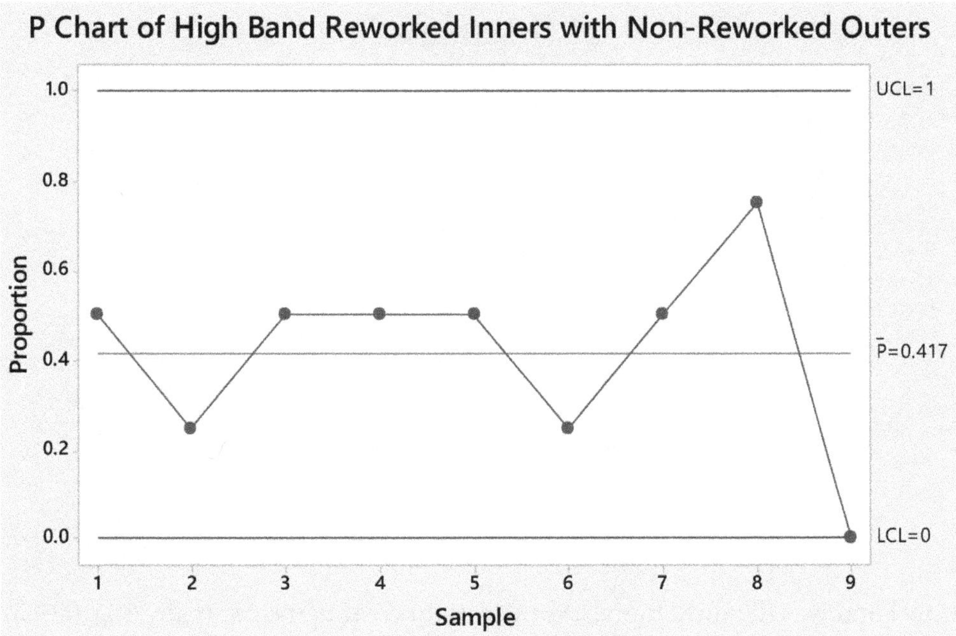

A p chart for the reworked outers with non-reworked inners was not created because all of them passed on the Waveometer so this chart was not needed.

Historical Data

The historical data provided by the manufacturing facility was data from the year 2016 on the scrap produced by the plant. The data was organized by the factory using an Excel spreadsheet. The specific data that was examined in this study was the highest scrap reasons based on quantity and the highest scrap reasons based on cost.

The histogram is useful tool that can be utilized in Six Sigma and Lean Six Sigma methodologies. Figure 13 is a histogram of the highest scrap reasons in cost. The x-axis is the cost in dollars and the y-axis is the reason code for the scrap. Figure 17 displays set up/reset as the top cost to the company and ID oversized as the second highest cost to the company. These two scrap reasons were not studied in this research. However, it is beneficial to the manufacturing facility to note the two reasons costing them the most in scrap.

Figure 17. Histogram of Highest Scrap Reasons (Dollar Amount). Data retrieved from historical data collected by manufacturing facility, March 29, 2017.

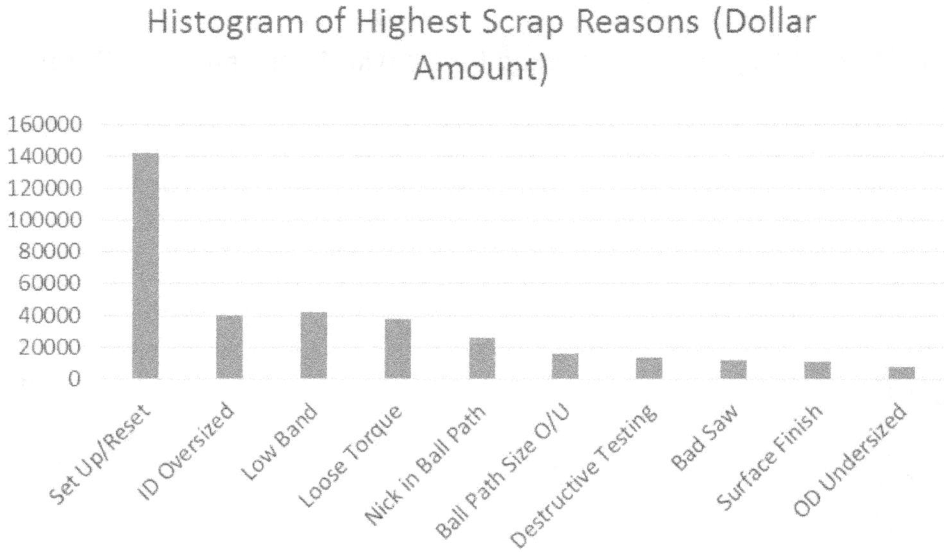

Low band was the third highest scrap reason costing the company $41,073 in the year 2016. Table 5 shows each reason codes cost to the company. Only the top 10 were examined as there are 117 scrap reason codes.

A Pareto chart can be used to organize the data by cumulative percentages. In Figure 14, "Other 1" is a combination of 4 different reason codes that all fell under mechanical defects. "Other 2" is a combination of 3 different

Table 5. Reason Codes and Dollar Amounts

Reason Code	Dollar Amount ($)
Set Up/Reset	$142,236
ID Oversized	$42,412
Low Band	$41,073
Loose Torque	$37,941
Nick in Ball Path	$26,853
Ball Path Size O/U	$16,760
Destruct6ive Testing	$14,123
Bad Saw	$11,956
Surface Finish	$11,760
OD Undersized	$8,520

reason codes that fell under cosmetic defects. Only the top 10 reason codes are shown below as there are 117 reason codes used. Figure 18 displays that low band was the top singular reason code used for scrap. The largest amount of pieces scrapped were due to low band. Low band, set up, and high band were responsible for 52.2% of the top 10 reasons. Set up was not studied in this research, but should be noted as accounting for the second largest quantity of scrap. Table 6 provides information about the highest scrap reasons, the quantities, the percentages, and the cumulative quantities and percentages.

A pie chart was created to analyze the low band and high band reason codes in relation to all of the others. There are two different types of low band codes. One is simply titled low band and the other is insert/noisy/low

Figure 18. Pareto Chart of Highest Scrap Reasons (Quantity Amount). Chart constructed in Excel software program by author. Data retrieved from historical data provided by the manufacturing facility. Retrieved March 29, 2017.

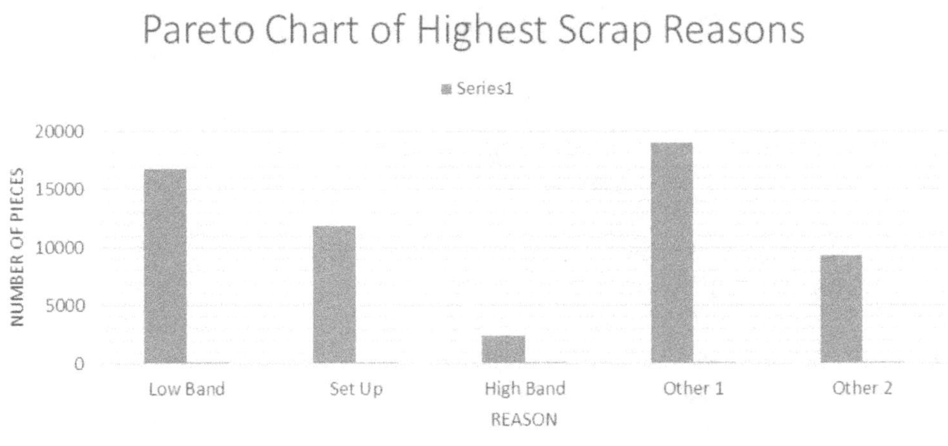

Table 6. Reason Codes and Quantities

Reason	Number of Pieces	Percent	Cum Num. of Pieces	Cum Percent
Low Band	16745	28.2	16745	28.2
Set Up	11855	20	28600	48.2
High Band	2390	4	30990	52.2
Other 1	18971	32	49961	84.2
Other 2	9323	15.7	59284	100

band. In the historical data, the insert/noisy/low band did not account for very much quantity or cost but was included in the pie chart. There are also two different types of high band codes. One is simply titled high band and the other is insert/noisy/high band. The insert/noisy/high band also did not account for very much in the overall quantity and cost from scrap, but was included in the pie chart.

Table 7 breaks down each section of the pie chart and gives their percentages. From this table, low band accounts for approximately 21% of the total scrap

Figure 19. Pie Chart of Reason Codes. Chart constructed in Minitab 17 software program by author. Data retrieved from historical data provided by the manufacturing facility. Retrieved March 29, 2017.

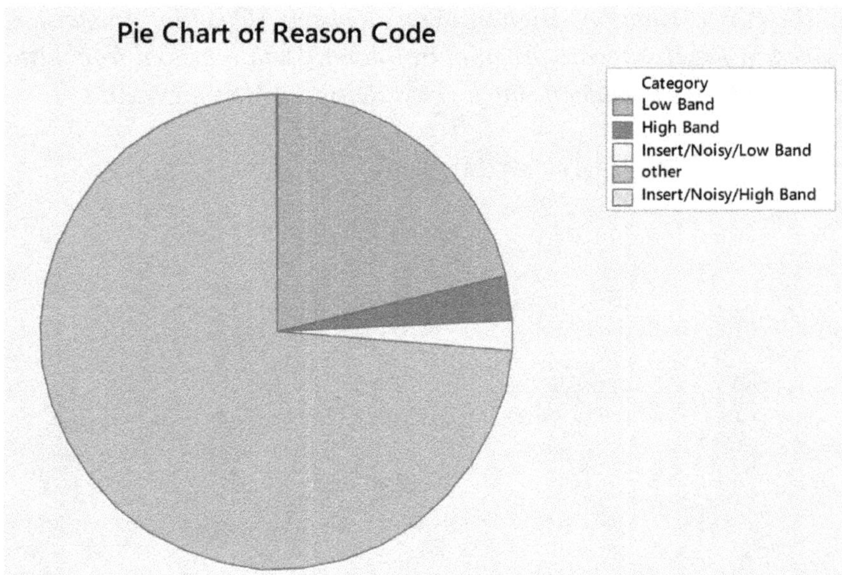

**For a more accurate representation see the electronic version.*

Table 7. Reason Codes and Percentages

Reason Code	Percent
Low Band	21.3%
High Band	3%
Insert/Noisy/Low Band	2%
Insert/Noisy/High Band	Insignificant
Other	73.7%

reasons. This is according to quantity, not cost. Insert/noisy/low band only makes up 2% of the total scrap reasons. 3% of the total scrap is due to high band defects. While insert/noisy/high band made up an insignificant amount of the total scrap. All of the other scrap reasons represent approximately 74% of the total scrap defect reasons.

CONCLUSION AND RECOMMENDATIONS

Quality improvement began with major inventions that spurred massive changes in industrial manufacturing. Quality assurance programs were being adopted by businesses by the 1970s in an effort to utilize quality improvement as a prevention method. The need to establish standards in the 1980s resulted in the creation of Total Quality Management, ISO 9000 (later 9001), Six Sigma, and Lean Six Sigma. The application of quality management within the manufacturing industry is still considered a recent development within the last fifty years.

The primary goal of this research was to identify the factors in the process for performance rework that were causing the most cost to the manufacturing company, measure and analyze the current process, and use these factors to make suggestions that would result in the reduction of cost and improvement of the process. The name of the facility was omitted from the research since the appropriate permission to publish the name was not granted. The research follows the Six Sigma and Lean Six Sigma method of thinking by using the DMAIC (defined, measured, analyzed, improved, and controlled) process.

Data for the research was collected at the facility in the month of March, 2017. Data was also used that the facility had previously collected from the year 2016 on their scrap details. The main limitation for the data collected by the author was that it used small sample sizes of 4 and a small lot size of 34 for one of the two scrap reasons studied. The factory where this research was conducted also had no previous data on rework fallout rates. Therefore, parts could still be non-conforming after being reworked. There was also no SPC currently used at the specific noise test inspection point.

- **Objective 1:** To appraise the existing Performance Rework Process at a local manufacturing company.

The existing Performance Rework Process at a local manufacturing company is a well-developed process. The only recommended change would come in the form of an adjustment to the destination of the rejected noise parts. Previously, the low and high band noise rejection parts were being sent to tear down and be reworked if they were over a certain quantity. If they were under a certain quantity, they were automatically scrapped. The parts rejected for low band noise should now all be sent to be scrapped. High band noise rejections can be sent to be reworked if they are over the specified quantity.

- **Objective 2:** To accurately measure noise levels of parts or parts processed, such as standard cylindrical series, RB (rubber mounted inserts), ER (non-spherical outer diameters), and SK (special order) Inners/Outers for performance ball bearings.

The noise levels are measured at the Anderon meter and the Waveometer. Both machines use a percentage. Anything 50% or less is pass or acceptable. Anything 51% or more is rejected. The process for measuring the noise levels of parts could be an area of additional research. A gage R&R study could be completed on both the Anderon meter and the Waveometer.

- **Objective 3:** To measure and analyze output of noise levels in hz, for variability using control charts in Minitab statistical software.

This is another area in which further study would be recommended. It is difficult to pinpoint exactly how many hertz a part is measuring as both the machines used to measure use percentages, not hertz. Charts for variability were not created during this study.

- **Objective 4:** To improve and control the Performance Rework Process. To develop and standardize sorting process to meet customer demand. To develop a more cost efficient process to meet customer demand.

The process for Performance Rework can be improved by taking the steps listed in this research. The process that the manufacturing facility follows was already a well thought out process. Minor adjustments to where the rejected parts are sent could save the company money and allows customer demands to be met in a timelier manner.

Following the use of pie charts, Pareto charts, and histograms, the given information displayed and found in the Excel spreadsheet in Appendix C, D, E, and F allows for multiple recommendations to be made to the manufacturing facility. From the 2016 data provided by the plant, it is obvious that low band comprises a large portion of their total scrap for both cost and quantity. In the past, the organization attempted to rework some of the low band in order to salvage the parts and save money. The facility had not collected any data on the fallout rate once the parts had been reworked and sent back through assembly. Therefore, they did not know if the parts they were reworking were still producing viable bearings. From the data collected on low band scrap, the final yield after rework was only 2.8%. Thus, the primary recommendation for this facility would be that all bearings that are rejected due to low band be scrapped. It is not worth the time of the hone operator and the money paid to that operator when the final yield for low band rework is so low. The factory would never break even.

As for high band, it was demonstrated by the data collected at the manufacturing factory, that after being reworked the inners did not do well when paired with non-reworked outers. However, the reworked outers all passed at the Anderon meter when they were paired with non-reworked inners. Therefore, the second recommendation is for the manufacturing facility to send bearings that are rejected for high band to be reworked on the hones and sent back through assembly. Since the reworked inners did not do as well as the reworked outers, it could be suggested that the factory only send the outers to be reworked and scrap the inners. This would need more research as only 34 high band bearings were studied in this thesis.

REFERENCES

Bellgran, M., & Säfsten, K. (2005). *Produktionsutveckling: utveckling och drift av produktionssystem*. Studentlitteratur.

Berardinelli, C. F. (2012, November). TO DMAIC or Not to DMAIC? *Quality Progress*, *45*(11), 72.

Bhamu, J., & Sangwan, K. S. (2014). Lean manufacturing: Literature review and research issues. *International Journal of Operations & Production Management*, *34*(7), 876–940. doi:10.1108/IJOPM-08-2012-0315

Bhuiyan, N., & Baghel, A. (2005). An overview of continuous improvement: From the past to the present. *Management Decision*, *43*(5), 761–771. doi:10.1108/00251740510597761

Breyfogle, F. W. (2003). *Implementing Six sigma: smarter solutions using statistical methods*. Hoboken, NJ: Wiley.

Broughton, R. (n.d.). *Flowchart shapes and description*. Retrieved April 9, 2017, from http://www.quality-assurance-solutions.com/flowchart-shapes.html

Brue, G., & Howes, R. (2006). *Six Sigma: The McGraw-Hill 36-hour's course*. New York: McGraw-Hill.

Chen, J. C., Li, Y., & Shady, B. D. (2010). From value stream mapping toward a lean/sigma continuous improvement process: An industrial case study. *International Journal of Production Research, 48*(4), 1069–1086. doi:10.1080/00207540802484911

Cut the Costs of Rework with Manufacturing Process Improvement. (n.d.). Retrieved April 9, 2017, from http://www.isapartners.org/business-education-resources/articles/cut-the-costs-of-rework-with-manufacturing-process-improvement

Demers, J. (2002). The Lean Philosophy. *CMA Management Magazine,* 32-33.

Ejderhov, C., & Åkerlund, A. (2015). *Rework Process: Determining the Current State of the Rework Process and Developing a New Process that Enables Transparent Rework at GE Healthcare UMEA* (Unpublished master's thesis). Luleå University of Technology.

John D. Dingell VA Medical Center, Detroit, Michigan. (n.d.). Retrieved April 10, 2017, from http://www.detroit.va.gov

Kumar, D. (2015). *Scrap Reduction in a Piston Manufacturing Industry: An Analysis Using Six Sigma and DMAIC Methodology*. Academic Press.

Lean Six Sigma Definition. (2014, September 12). Retrieved April 9, 2017, from http://www.investopedia.com/terms/l/lean-six-sigma.asp

Quality America Inc. (2013, January 1). Retrieved April 9, 2017, from http://qualityamerica.com/LSS-Knowledge-Center/statisticalprocesscontrol/statistical_process_control_defined.php

Quality Management. (n.d.). Retrieved April 9, 2017, from http://www.businessdictionary.com/definition/quality-management.html

Salah, S., Rahim, A., & Carretero, J. (2010). The integration of Six Sigma and lean management. *International Journal of Lean Six Sigma, 1*(3), 249–274. doi:10.1108/20401461011075035

Taylor Hobson Centre of Excellence. (n.d.). *Roundness Filters and Harmonics* [PowerPoint Slides]. Retrieved April 9, 2017, from http://www.taylorhobsonserviceusa.com/uploads/2/5/7/5/25756172/roundness_filters.pdf

The Define Measure Analyze Improve Control (DMAIC) Process. (n.d.). Retrieved April 9, 2017, from http://asq.org/learn-about-quality/six-sigma/overview/dmaic.html

Total Quality Management. (n.d.). Retrieved April 9, 2017, from http://asq.org/learn-about- quality/total-quality-management/overview/overview.html

Valles, A., Sanchez, J., Noriega, S., & Gómez Nuñez, B. (2009). Implementation of Six Sigma in a Manufacturing Process: A Case Study. *International Journal of Industrial Engineering*, *16*(3), 171–181.

What I. S. S. (n.d.). Retrieved April 9, 2017, from http://www.isixsigma.com/new-to-six- sigma/getting-started/what-six-sigma

What Are Attributes Control Charts? (n.d.). Retrieved April 9, 2017, from http://www.itl.nist.gov/div898/handbook/pmc/section3/pmc33.htm

Womack, J. P., & Jones, D. T. (2003). *Lean thinking: banish waste and create wealth in your corporation*. New York: Free Press.

APPENDIX A

Flow Chart Shapes and Definitions

Figure 20. Flowchart shapes and description
Retrieved April 9, 2017, from http://www.quality-assurance-solutions.com/flowchart-shapes.html.

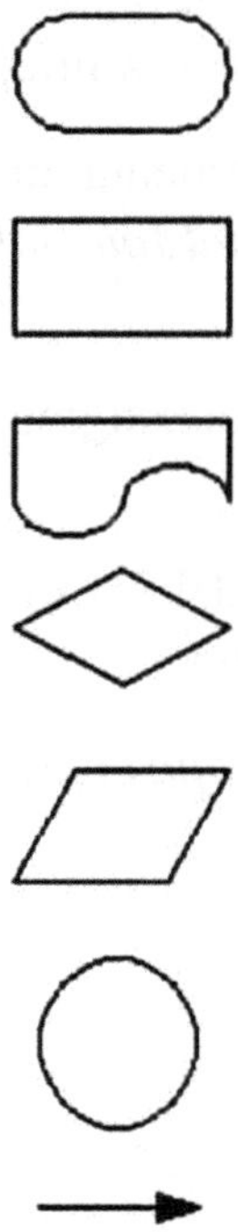

Start/End
The terminator symbol marks the starting or ending point of the system. It usually contains the word "Start" or "End."

Action or Process
A box can represent a single step ("add two cups of flour"), or and entire sub-process ("make bread") within a larger process.

Document
A printed document or report.

Decision
A decision or branching point. Lines representing different decisions emerge from different points of the diamond.

Input/Output
Represents material or information entering or leaving the system, such as customer order (input) or a product (output).

Connector
Indicates that the flow continues where a matching symbol (containing the same letter) has been placed.

Flow Line
Lines indicate the sequence of steps and the direction of flow.

APPENDIX B

Value Stream Map Shapes and Definitions

Figure 21.
Retrieved from: John D. Dingell VA Medical Center, Detroit, Michigan. (n.d.). Retrieved April 10, 2017, from http://www.detroit.va.gov.

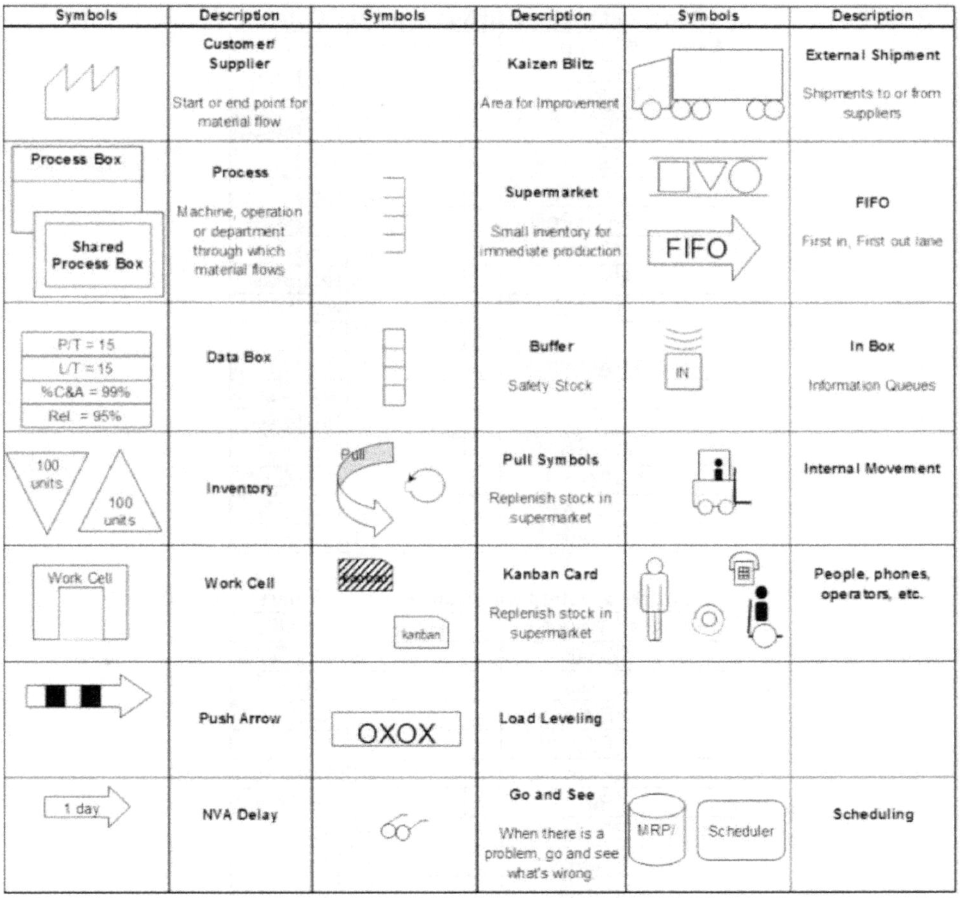

Symbols	Description	Symbols	Description	Symbols	Description
	Customer/ Supplier Start or end point for material flow		**Kaizen Blitz** Area for Improvement		**External Shipment** Shipments to or from suppliers
Process Box Shared Process Box	**Process** Machine, operation or department through which material flows		**Supermarket** Small inventory for immediate production	FIFO	**FIFO** First in, First out lane
P/T = 15 L/T = 15 %C&A = 99% Rel = 95%	**Data Box**		**Buffer** Safety Stock	IN	**In Box** Information Queues
100 units 100 units	**Inventory**	Pull	**Pull Symbols** Replenish stock in supermarket		**Internal Movement**
Work Cell	**Work Cell**	kanban	**Kanban Card** Replenish stock in supermarket		**People, phones, operators, etc.**
	Push Arrow	OXOX	**Load Leveling**		
1 day	**NVA Delay**		**Go and See** When there is a problem, go and see what's wrong	MRP/ Scheduler	**Scheduling**

APPENDIX C

Excel Spreadsheet of High Band Reworked Inners and Outers

Figure 22. Excel spreadsheet created by author, April 2017

Reworked Inners with Non-reworked Outers				
Sample #	Pass/Fail		Sample #	Pass/Fail
1	Pass		18	Not Pass
2	Not Pass		19	Pass
3	Pass		20	Not Pass
4	Not Pass		21	Pass
5	Pass		22	Pass
6	Not Pass		23	Not Pass
7	Pass		24	Pass
8	Pass		25	Not Pass
9	Not Pass		26	Not Pass
10	Pass		27	Pass
11	Not Pass		28	Pass
12	Pass		29	Not Pass
13	Pass		30	Not Pass
14	Not Pass		31	Pass
15	Pass		32	Not Pass
16	Not Pass		33	Pass
17	Pass		34	Pass

Reworked Outers with Non-reworked Inners				
Sample #	Pass/Fail		Sample #	Pass/Fail
1	Pass		18	Pass
2	Pass		19	Pass
3	Pass		20	Pass
4	Pass		21	Pass
5	Pass		22	Pass
6	Pass		23	Pass
7	Pass		24	Pass
8	Pass		25	Pass
9	Pass		26	Pass
10	Pass		27	Pass
11	Pass		28	Pass
12	Pass		29	Pass
13	Pass		30	Pass
14	Pass		31	Pass
15	Pass		32	Pass
16	Pass		33	Pass
17	Pass		34	Pass

APPENDIX D

Excel Spreadsheet of Low Band Reworked Inners

Figure 23. Excel spreadsheet created by author, April 2017

Low Band Reworked Inners

Sample #	Pass/Fail	Sample #	Pass/Fail	Sample #	Pass/Fail	Sample #	Pass/Fail	Sample #	Pass/Fail	Sample #	Pass/Fail
1	Pass	19	Not Pass	37	Not Pass	55	Not Pass	73	Pass	91	Pass
2	Not Pass	20	Not Pass	38	Pass	56	Not Pass	74	Not Pass	92	Pass
3	Not Pass	21	Not Pass	39	Not Pass	57	Pass	75	Not Pass	93	Pass
4	Pass	22	Not Pass	40	Not Pass	58	Pass	76	Pass	94	Pass
5	Not Pass	23	Pass	41	Pass	59	Pass	77	Not Pass	95	Pass
6	Pass	24	Pass	42	Not Pass	60	Not Pass	78	Not Pass	96	Not Pass
7	Not Pass	25	Not Pass	43	Pass	61	Not Pass	79	Not Pass	97	Not Pass
8	Not Pass	26	Not Pass	44	Not Pass	62	Pass	80	Not Pass	98	Not Pass
9	Not Pass	27	Not Pass	45	Not Pass	63	Not Pass	81	Pass	99	Not Pass
10	Pass	28	Not Pass	46	Not Pass	64	Not Pass	82	Not Pass	100	Pass
11	Not Pass	29	Pass	47	Pass	65	Pass	83	Not Pass	101	Not Pass
12	Not Pass	30	Pass	48	Not Pass	66	Not Pass	84	Not Pass	102	Not Pass
13	Not Pass	31	Not Pass	49	Not Pass	67	Pass	85	Pass	103	Pass
14	Pass	32	Not Pass	50	Pass	68	Not Pass	86	Not Pass	104	Not Pass
15	Not Pass	33	Pass	51	Not Pass	69	Not Pass	87	Pass	105	Pass
16	Not Pass	34	Not Pass	52	Not Pass	70	Pass	88	Not Pass	106	Pass
17	Pass	35	Pass	53	Pass	71	Not Pass	89	Not Pass		
18	Not Pass	36	Not Pass	54	Pass	72	Not Pass	90	Not Pass		

APPENDIX E

Excel Spreadsheet of Low Band Reworked Outers

Figure 24. Excel spreadsheet created by author, April 2017

Low Band Reworked Outers

Sample #	Pass/Fail	Sample #	Pass/Fail	Sample #	Pass/Fail	Sample #	Pass/Fail	Sample #	Pass/Fail	Sample #	Pass/Fail
1	Not Pass	19	Not Pass	37	Pass	55	Not Pass	73	Not Pass	91	Not Pass
2	Not Pass	20	Not Pass	38	Not Pass	56	Not Pass	74	Not Pass	92	Not Pass
3	Pass	21	Not Pass	39	Not Pass	57	Not Pass	75	Not Pass	93	Pass
4	Not Pass	22	Not Pass	40	Not Pass	58	Not Pass	76	Not Pass	94	Not Pass
5	Not Pass	23	Not Pass	41	Not Pass	59	Not Pass	77	Not Pass	95	Not Pass
6	Not Pass	24	Not Pass	42	Not Pass	60	Not Pass	78	Not Pass	96	Not Pass
7	Not Pass	25	Not Pass	43	Not Pass	61	Not Pass	79	Not Pass	97	Not Pass
8	Pass	26	Not Pass	44	Not Pass	62	Not Pass	80	Pass	98	Not Pass
9	Not Pass	27	Pass	45	Not Pass	63	Not Pass	81	Not Pass	99	Not Pass
10	Not Pass	28	Not Pass	46	Not Pass	64	Pass	82	Not Pass	100	Not Pass
11	Not Pass	29	Not Pass	47	Not Pass	65	Not Pass	83	Not Pass	101	Not Pass
12	Not Pass	30	Not Pass	48	Pass	66	Not Pass	84	Not Pass	102	Not Pass
13	Not Pass	31	Not Pass	49	Pass	67	Not Pass	85	Not Pass	103	Not Pass
14	Not Pass	32	Not Pass	50	Not Pass	68	Not Pass	86	Pass	104	Pass
15	Not Pass	33	Not Pass	51	Not Pass	69	Not Pass	87	Not Pass	105	Not Pass
16	Pass	34	Not Pass	52	Not Pass	70	Not Pass	88	Not Pass	106	Not Pass
17	Not Pass	35	Pass	53	Not Pass	71	Not Pass	89	Not Pass		
18	Not Pass	36	Not Pass	54	Not Pass	72	Not Pass	90	Not Pass		

APPENDIX F

Excel Spreadsheet of Low Band Reworked Inners and Outers

Figure 25. Excel spreadsheet created by author, April 2017

Reworked Inners and Outers	
Sample #	Pass/Fail
1	Not Pass
2	Pass
3	Not Pass
4	Not Pass
5	Not Pass
6	Pass
7	Not Pass
8	Pass
9	Not Pass
10	Not Pass
11	Not Pass
12	Not Pass
13	Not Pass

s

Chapter 2
Lean Tools to Evaluate Lighting Conditions

ABSTRACT

Lighting in indoor and outdoor settings has a purpose of providing an environment that is deemed as being comfortable and safe with the highest amount of light at the lowest possible consumption of energy. On the downside, there are instances where lighting is not as efficient as it is expected to be. For instance, the current outdoor lighting infrastructure at the University Campus (Referred to as "Campus") located in a southern town in the United States of America proves this fact. In reality, the lighting on this campus is outdated when compared to the expected lighting standards. The main objective of the chapter is to use lean tools to analyze data on the light luminosity, pole locations, and other conditions. The results suggest a recommendation for a lean solution for the implementation of new and additional lighting on the Campus, which is deemed to provide efficiency and safety.

INTRODUCTION

Lean is a concept that offers a competitive edge to manufacturers as far as reducing costs, increasing productivity, and quality is concerned (Bhamu and Singh, 2014). One definition that augurs well with the present study in this chapter is that it is considered to be a strategic tool that plays a fundamental role in the identification and the provision of solutions to the problems in an organization (Atkinson, 2010). Lean tools have evolved from the time

DOI: 10.4018/978-1-5225-4062-5.ch002

they were used in the Toyota Production System, in Japan's automotive sector (Becker, 1998). Torielli et al. (2011) consider Lean to be a critical tool when it comes to the reduction of overall costs in an organization. As a matter of fact, the Lean philosophy is linked to waste reduction, and it was not until recently that that these tools were noted to be of vital importance in the promotion of environmental sustainability and efficiency (Torielli et al., 2011). It is; therefore, important to note that lean is a mindset that can be integrated in the organization, thus instigating the essence of sustainability and efficiency, in line with the view by Torielli et al. (2011).

With the above insight on the role that lean tools and the lean philosophy therein play at the organization level, it is essential to note that these elements can be used to assess and address the problem of poor and obsolete lighting on the paths near buildings on the Campus. It is notable that compromised safety is a problem that is often attributed to poor lighting. Given that improved lighting is evidenced by Clarke (2008) to be an effective means of promoting safety, recommendations that address this issue will have to be identified. With the knowledge that improving lighting on the Campus will come at a cost, one way that the costs can be reduced is through the use of strategies that foster waste reduction, and promote quality and efficiency using lean tools. Additionally, with the insight that the combination of lean tools in organizations plays a fundamental role in ensuring that quality, sustainability and efficiency is achieved, the present study will also consider that aspect of problem. This is especially true with the fact that lean tools reinforce each other, as established by Clarke (2008), and so combining a few in this study is quite in order.

Objective of the Study

The main aim for the study is to determine the level of light wastage and pollution at all locations of the campus, emanating from poor lighting conditions. The results will then be used to make recommendations that revolve around lean principles that the university should consider adopting, in the bid to promote energy efficiency, reduce wastage and light pollution, minimize costs, and improve quality of lighting. The research questions that the study will seek to answer are the level of current light pollution at all locations on the Campus and the recommendations for modification or enhancement the exterior lighting of Campus?

LITERATURE REVIEW

The review of literature attempts to critically look at certain themes that are essential to address when looking at the state of energy costs that is attributed to lighting, and what can be done to foster efficiency.

Energy Costs

It is essential to note that buildings consume considerable amounts of energy usage, which was identified to be at a rate of between 20% and 40% (Krarti, Erickson, & Hillman, 2005). The figure would probably rise when adding the outdoor lights that are located around buildings. Unfortunately, out of the energy that goes into lighting, an estimate of about 30% is wasted energy, as noted by Delaney, O'Hare, and Ruzzelli (2009).

Energy Awareness

Given the fact that high energy costs are linked to lighting systems, management systems have been used to save energy, thus promoting energy efficiency. As a matter of fact, in some modern buildings that are energy aware, as identified by Delaney, O'Hare, and Ruzzelli (2009) tend to have lighting systems that maximize on the essence of energy efficiency.

Lighting Systems That Are in Use

The examples of lighting mechanisms that is in place include artificial lighting. When speaking of artificial lighting, it is essential to note that some of the systems that are already in use are inefficient. For instance, as established by Delaney, O'Hare, and Ruzzelli (2009), most systems use sensors that are fitted with actuators, which determine turning on/off mechanisms, or the dimming of lights when necessary. Standard tuning values that are obtained from statistical calculations are employed in such sensors, which lack the optimization to sense the environment. The aftermath of using such like sensors is the presence of energy wastage and overall inefficiencies.

Going the Lean Way

It is essential to note that many ideas can be explored in lean. Surveying through the various strategies that can be employed is essential, in order to determine the best option that can be adopted.

Visser (2014) identified that often, lean tends to be practice or tool-based. Reviews have shown that some of the most common lean tools that are adopted by companies include; continuous improvement programs, continuous flow production, just in time production, quick changeovers, and the Kanban systems (Visser, 2014). Visser (2014) further established that lean can be described as per certain critical determinants such as the actions that are taken by a company, the strategies implemented, and the benefits that are felt as a result. For instance, the determinants can take note of waste elimination, continuous improvement, the presence of multi-functional teams, zero defects, decentralization of teams, vertical information systems, production based on pull systems, and JIT (Visser, 2014).

In the study by Visser (2014), lean was classified into four categories namely; JIT, Total Quality Management, Total Productive Maintenance, and Human Resource Management.

Just-In-Time (JIT)

JIT is concerned with cutting down on wastes and eliminating unnecessary delays, by the identification of hidden problems in a given process (Javadian Kootanaee, Babu, & Talari, 2013). Here, autonomation comes in handy. A study by Boakye-Adjei, Thamma, and Kirby (2014) shed light on Autonomation (jidoka). This strategy allows frequent monitoring of various sections, thus facilitating the detection of quality issues that occur. Autonomation instigates the usage of complex systems, which automate a particular process, thus allowing monitoring and control, promoting a mechanism of self-diagnosis (Boakye-Adjei, Thamma, & Kirby, 2014). See Figure 1, which shows the overall concept of autonomation.

From Figure 1, it is apparent that the role that autonomation plays is substantial to the promotion of continuous improvement.

Figure 1.

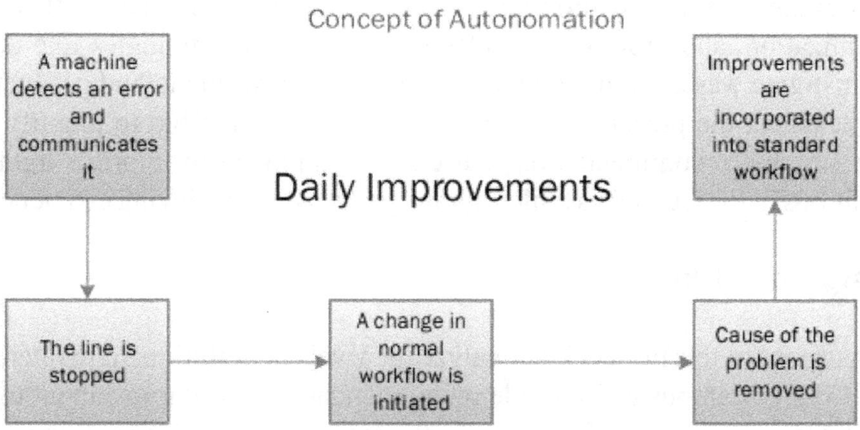

Total Quality Management

Total Quality Management (TQM) has its focal point on continuous improvement by improving the existing processes, with a reduction in the amount of resources that are employed (Andersson, Eriksson, & Torstensson, 2006). Here, value stream mapping comes in handy (Visser, 2014). When narrowing down to VSM, it is essential to note that this lean tool is concerned with process improvement, with the development of a culture that fosters problem-solving, continuous growth, and improvement. In the study by Kurdve et al. (2011), VMS was noted to be a fundamental aspect in that provides a comprehensive analysis of the main processes and the critiques therein. The study revealed that environmental benefits were likely to emerge from VMS, in the form of less energy use, reduced emission, and a reduction of waste (Kurdve et al., 2011).

Total Productive Maintenance (TPM), on the other hand, is concerned with preventive and predictive maintenance strategies, including maintenance optimization, with the aim of maximizing equipment effectiveness, thus promoting the improvement of maintenance practices (Marquez & Gupta, 2006). Finally, Human Resource Management (HRM) fosters cultural empowerment, where employees own their processes (Visser, 2014). It is essential to note that studies that are concerned with the area of cost-effectiveness in energy consumption and usage identify that about 20%-40% energy saving can be achieved in and near buildings, with the potential for more saving if the right systems are in place (Biggart & Lutzenhiser, 2007).

The significance of the literature confirms that lean tools can be applied to an organization, thus promoting benefits that are concerned with waste minimization, and promotion of efficiency. With lean strategies like those that instigate waste minimization, continuous improvement, and productive maintenance, the present study can be propelled, in the bid to identify the most suitable recommendation that can be employed to improve lighting conditions in the Campus's buildings and paths, while achieving efficiencies.

Design of Study

The purpose of this project is to analyze the levels of nighttime light intensity on a College campus and provide recommendations for improvement using lean tools and principles. Specifically, the main objective is identifying areas that could be classified as dark and/or unlit. In addition, our study outlined sources of light pollution on campus or excessive light intensity.

Autonomation and Its Application

It should be noted that in the data collection of the light measurements and pole locations, the lean tool dubbed as autonomation (*jidoka*) was employed. The application of autonomation was pertinent in the processes of measuring and mapping of light intensity and poles. Autonomation (automation with the presence of human element), as identified by Subramaniam et al. (2009), is a principle that is concerned with stopping work the moment a defect is determined. As noted by Subramaniam et al. (2009), autonomation involves the execution of four steps, which were applied in this paper, which include;

- Detection of abnormalities
- Stopping the process
- Fixing the problem
- Identifying a corrective action

The above steps were vital in the method of data collection, where the following was done;

1. The discovery of anomalies in the lighting conditions at the selected areas of the Campus prompted data collection and documentation based on the findings obtained from various areas of the Campus,

2. This was followed by "stopping" the series of the anomaly to assess the lighting problem that was present on the campus,
3. Step two was followed by fixing the lighting problem, and
4. Identification and addressing the root cause therein by determining the necessary corrective actions to cater for the lighting problems on the Campus, which are discussed in this chapter, and recommendations outlined as well.

It is worth to consider that the role of autonomation is to remove judgment and it is aimed at using humans to address the problems that emerge, which have been identified by machines. This concept is apparent in the study with the use of light meters, lux meters, and GPS technologies to assess the lighting problems on the Campus. The results yield the need to take human action to address the identified issues, through the implementation of lasting corrective actions. The use of the light meters, lux meters, and GPS coordinates confirm the fact that these tools added value to the data collection and assessment.

Measuring Light Luminosity

To collect data for light luminosity light meters were used to measure light intensity in units of lumens. The entire campus was walked with a lux meter, illustrated in Figure 2, to measure the light intensity.

Figure 2. Lux Meter

The light meter has a tiny bone shape with lightweight and a small size case design that is suitable for handling with one hand. The hand cord design provides extra protection to the instrument, in particular for the user's one hand operation. Exclusive photo diode and color-correction filter light sensor, spectrum meets C.I.E. photopic. The units are Lux and Feet-candle with an auto range of 0 to 20,000 Lux. The device is built in microprocessor circuit that assures excellent performance and accuracy. Concise and compact buttons arrangement provides smooth operation with the capacity to memorize the maximum and minimum value with recall with no adjustment needed. There is also a hold function to freeze the current reading values with ease (Horm Hardware, 2017)

Readings from each pole are recorded approximately three times to ensure accuracy. The data collection devices that were employed in the study were calibrated to a properly working state before each work night. In addition to collecting the light intensity, data such as the light pole identification number, type, height, and miscellaneous data such as dual lamps and proximity to buildings were recorded. GPS data were also recorded with a single handheld GPS device for each pole. Poles that contain more than one lamp were recorded twice. At the end of each workday, all data were downloaded from the devices into a spreadsheet file within Microsoft Excel for easy viewing and to clear space on the data collection devices for the next collection session. Table 1 provides a snapshot of the data collected near a campus building.

Table 1. Snapshot of Data collected near HALL #1

Pole Type	Pole Height	GPS North	GPS West	D1 Proximity	D1 Light Intensity	D2 Proximity	D2 Light Intensity	D3 Proximity	D3 Light Intensity	Miscel-laneous
STEEL	362	37° 14.615'	77° 25.058'	91	89.73	138	22.66	125	67.42	HALL # 1
STEEL	356	37° 14.601'	77° 25.065'	372	17.41	553	7.66	498	2.93	HALL # 1
STEEL	356	37° 14.609'	77° 25.067'	497	8.68	568	10.47	286	17.85	HALL # 1
STEEL	358	37° 14.621'	77° 25.065'	453	17.57	637	6.78	626	10.37	HALL # 1
STEEL	351	37° 14.610'	77° 25.111'	562	12.45	369	15.27	666	13.42	HALL # 1
STEEL	355	37° 14.619'	77° 25.112'	406	6.53	407	10.83	543	5.45	HALL # 1
STEEL	353	37° 14.613'	77° 25..135'	593	8.25	307	13.57	567	4.48	HALL # 1

After the campus surroundings were surveyed, the information was integrated into a Geographic Information Systems (GIS) software, ArcGIS 9.3. ArcGIS is a system that allows a user who relies on accurate geographical information to make decisions. It facilitates collaboration to easily author data, maps, globes, and models to view, understand, question, interpret, and visualize data in many ways that reveal relationships, patterns, and trends in the form of maps, globes, reports, and charts. ArcGIS is used for the project to visually see the actual locations of the lights upon data collection as well as to visually see areas of the campus that lack lighting. ArcGIS is used to analyze and query specific relationships between areas where appropriate lighting standards are met and areas where there are deficiencies. The software allows designing of complex queries to understand the lighting of the campus beyond the passive observations. Once in the GIS environment, each pole and their corresponding information would be presented. Aerial photography and imaging allowed for visual representation of the Campus with the surveyed pole location layered on top. Figure 3 provides an aerial image of the Campus prior to data collection.

METHODOLOGY

The main concern of this study was to identify how to measure lighting intensity and how to map it, how light conditions could be recorded, accurately, the technology that would be used, and how the areas that were in need of improvement could be identified.

To provide a benchmark for determining lighting conditions, the study utilized critical pointers, which were drawn from the research conducted by Robert Samuels (1995), on crime and lighting conditions. Samuels concluded in his study that only 3% of crimes investigated on college campuses occurred in lighting conditions of 20 Lux or more while 40% of all crimes investigated in his studies occurred in low lighting conditions of 5 Lux or less. From this research, the benchmark was set on identifying all areas on campus with lighting conditions under 5 Lux and over 20 Lux. Although most crimes committed against university students are most likely to occur off campus, protecting students, faculty, staff, and visitors while on campus is an important priority as well. While lighting is an effective way to reduce crime, it may not eliminate all forms of criminal activity. A study conducted by Farrington and Welsh (2004) found that improved lighting can reduce crime by nearly

Figure 3. Aerial Map of the campus

20% and should be included as an element of a crime reduction program and can be coordinated with efforts of a police enforcement program.

With regards to light pollution, there is a growing concern when it comes to excessive lighting. This is most relevant in major cities and industrial areas as many developments expand and there is high demand for artificial lighting. As part of this study, the identification of light pollution is also important as any light pollution could be identified to promote a reduction in energy costs for the University.

Field Work and Mapping

To complete the field work portion of the project, using a light meter to check the light intensity, a walk around the whole campus was done. The overall objective was to inventory and measure light intensity for each pole (outside light sources). Three student interns used two light meters to record light intensity for each location with five readings to ensure an accurate measurement was achieved. In addition to recording the intensity data such as the pole number, GPS coordinates via Garmin Dakota 20 GPS device capable of holding up to 1,000 waypoints, proximity, pole type, pole height, and placement relative to a university building or parking lot were recorded. The aerial image and GIS data allowed for an aerial display of the campus showing locations for all buildings with accurate GIS data for street names, street widths, and street lengths. After all, measurements were taken they were then mapped in a GIS environment using ArcGIS 9.3 software. The resulting map can be found in Figure 4. A total of 807 existing lights were inventoried.

Measuring Lighting Conditions

When measuring the lighting conditions, in general, several units of measurement can be used, but the most common measurement unit is the lux. The lux is the measure of light intensity. Lumens is also a measure of light intensity. A lumen is the total light output from a source. The lumen quantifies the amount of light energy produced by a lamp at the lamp, not by the energy input, which is indicated by the wattage or horsepower. For example, a 75-watt incandescent lamp can produce 1000 lumens while a 70-watt high-pressure sodium lamp produces 6000 lumens. Lux is a metric unit that indicates the density of light that falls on a surface. It is equal to one lumen per square meter. Lux readings are used to measure the adequacy of lighting and can only give information regarding the light intensity in a particular area. The lux readings do not provide anything with regards to a comparison of equipment used (the type of light bulb or lighting structure). Natural sunlight measures 50,000 lux, typical indoor lighting ranges from 100 – 1,000 lux, and moonlight measures 1 lux.

Figure 4. An Aerial Map of Collected Data

The Characteristics of Lighting Types

With regards to lighting, the most common types of lighting found on many streets and parking lots constitute High-Pressure Sodium (HPS), Metal Halide (HD) and Light Emitting Diode (LED) lights. Although MH and HPS lights dominate the exterior lighting industry, LED, lights are beginning to gain ground. MH and HPS are two types of high-intensity discharge lights that give a brighter effect to an area and are widely used for outdoor lighting. LED lights generate less heat than MH and HPS, but look brighter in most applications as they have a wider range when it comes to color rendering. While the MH bulbs produce a blue-white color, HPS bulbs produce amber, white lights. With the same electricity, MH produces less rated lumens than HPS lamps while LED produced far much less when compared to MH and HPS. When looking at the glow that MH lights produce, it is clear that more sky glow (light pollution) is apparent compared to the HPS. LED practically produces little to no sky glow. MH lamps have been found to be four to five times more efficient than HPS lamps and are considered to be the preferred lighting choice due to the white light environment. LED lights, just like MH and HPS lamps, are also efficient. However many issues arise due to their operation in low-temperature environments. When looking at the HPS lamps, it is apparent that the large color display from these lights is useful in areas where a color reduction is not critical. Some advantages that LED lighting has over MH, and HPS lighting is that it does not include mercury, lead, or other harmful hazards, and a quick run-up time. With regards to cost, LED lighting currently costs more to purchase than traditional MH and HPS fixtures. However, the reduction in replacing LED lights and potential power savings may reduce the overall life-cycle cost when compared to MH and HPS.

ANALYSIS AND RESULTS

The results of the analysis conducted on the data collected with recommendations for improving lighting conditions on campus a statistical analysis as well as a life cycle analysis is conducted.

T- Test and Analysis of Variance

A t-test was conducted to assess whether the recommended normal light intensity level of the commercial building and parking of 5 lux could, in fact, be proven on the Campus. The results showed that the P-value = 0.000. This indicates that the light intensity on the campus, in general, can be measured up to the national standard found in the literature. The detail is illustrated Figure 5.

Next, an analysis of variance (ANOVA) was conducted to assess if there was a significant difference in the light exterior intensity for buildings on the campus. The results below showed that the P-value to be 0.000.

Figure 5. Figure 2 T-Test Summary

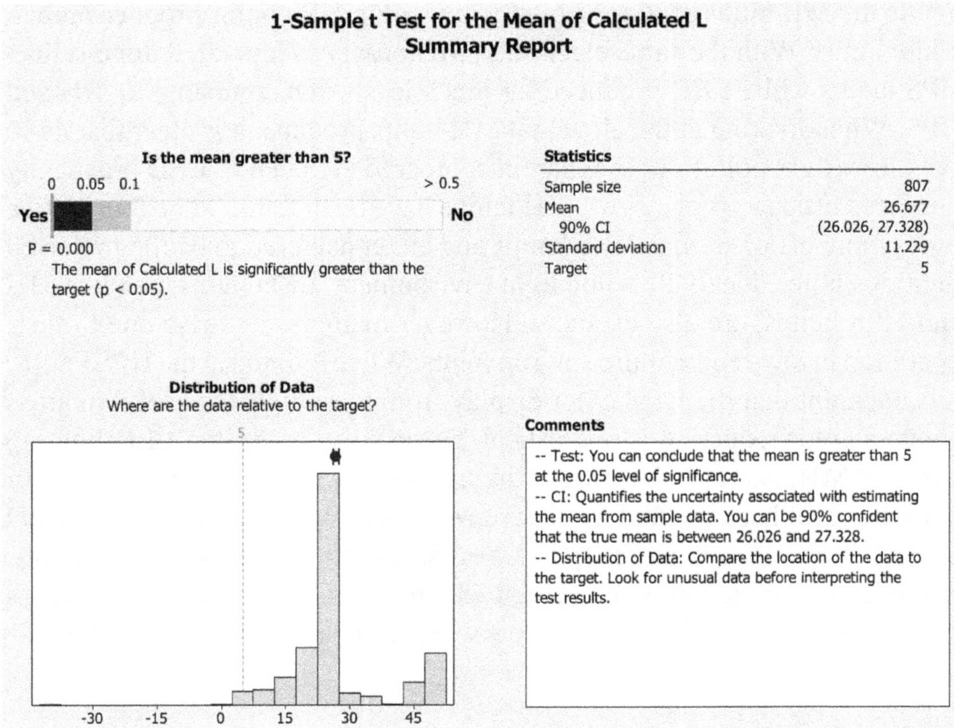

Table 2.

Source	DF	SS	MS	F	P
Regression	4	259159	64790	7.11	**0.000**
Residual Error	796	7258150	9118		
Total	800	7517309			

The results indicated that the difference between the light intensity among buildings was significant and, therefore, necessary attention was needed. A One-way ANOVA was also conducted to assess if there will be a significant difference in the light intensity between all buildings on campus. As illustrated below, the results indicated a P-value = 0.000 confirming a significant difference in the light intensity.

Table 3.

Source	DF	SS	MS	F	P
Light intensity	55	53688.0	976.1	15.29	**0.000**
Error	751	47947.1	63.8		
Total	806	101635.1			

Regression Analysis

A regression analysis was used to model and analyze the relationship between a light intensity and the independent variables such as baseline, pole height, proximity and time of collection. More specifically, the technique helped to understand how light intensity changes when any one of the independent variables is varied, while the other independent variables are held fixed. A regression model relates Y to a function of $\mathbf{X}$ and $\boldsymbol{\beta}$.

$$Y \approx f(X, B)$$

With:

B = unknown parameters; this may be a scalar or a vector

X = independent variables

Y = dependent variable

In this case, for prediction and forecasting purposes, the regression function is provided below. Based on the data collected the regression function is provided below.

Light Intensity = 53.3 - 10.8 Baseline Measurement + 20.0 Baseline Time - 0.165 Pole Height + 0.00089 Proximity

This equation was used for the calculated light intensity for every location on the Campus. The results are shown in Figure 6.

Figure 6. Pareto Chart of Building Relativity

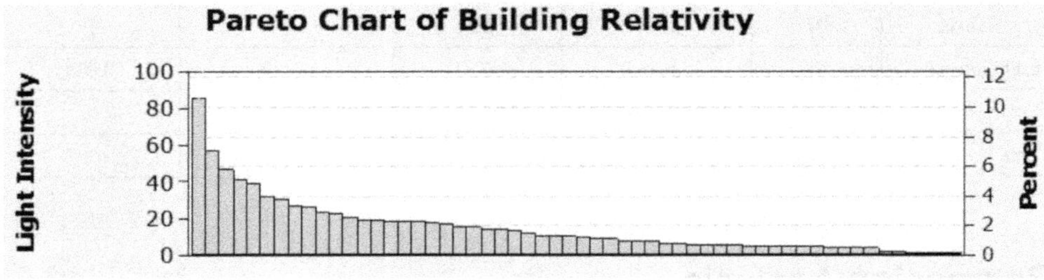

Pareto Analysis

The Pareto principle (also known as the 80-20 rule, the law of the vital few, and the principle of factor sparsity) states that, for many events, roughly 80% of the effects come from 20% of the causes.

Business-management consultant Joseph M. Juran suggested the principle and named it after Italian economist Vilfredo Pareto, who observed in 1906 that 80% of the land in Italy was owned by 20% of the population; he developed the principle by observing that 20% of the pea pods in his garden contained 80% of the peas.

In this case, we went to see what the buildings are making up the 80% of the lighting of the campus. This will help us to prioritize in making recommendations for modification, enhancement or addition of exterior lighting for the Campus. We can recommend based on the results shown in the graph below illustrated that any buildings above four (4) percent would not be part of our recommendation for attention in the next 20 years.

Based on the analysis provided in Figure 7 we recommend that any building above four (4) percent not to be included in the 20 years plan to fix lighting issues on campus. As examples, according to the chart above, the Campus streets are the most lit areas on campus. This explains why most of the lighting complaints were not for the streets; instead, they were for buildings and parking lots. These areas are also areas where crime is most likely to occur. Furthermore, Hall #2 is the next best location on campus for sufficient lighting. This can be explained by the fact that it is one of the more recent buildings constructed on the campus where current lighting technologies were implemented.

A surprise came from some of the newest buildings that were noted to have very low light intensity such as Hall #1. Our analysis indicated that as the height of the light pole increased light intensity decreased. Table 4 provides the

Figure 7. Pareto Chart of Calculated Light Intensity

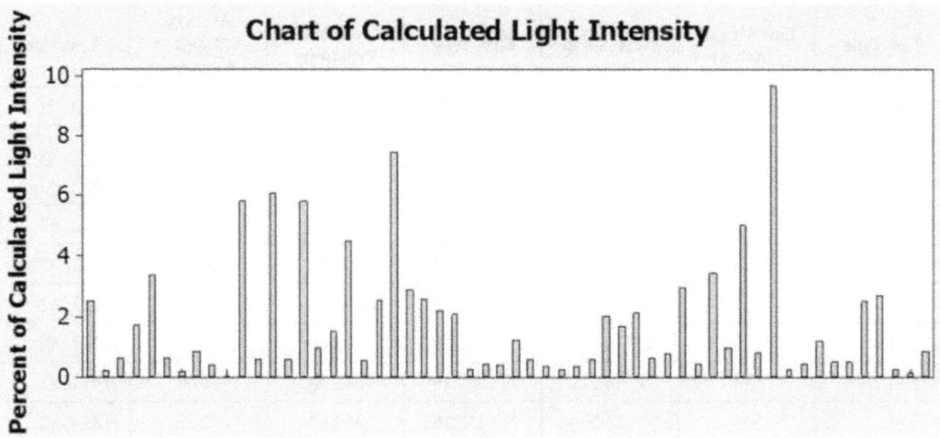

results of the data collection effort at both Hall #2 and Hall #1. Furthermore, the lights on some poles faced downward (straight down/downward angle), causing a reduction in luminosity of the lighted areas. Therefore, most of the lights that were placed at a downward angle were covering a wider range of area with the least intensity compared to the ones facing straight down. The data provided suggests that the majority of the taller lights were subjected to light pollution Based on our observation, we recommend having the light poles at a maximum height of 400 inches or ±30 feet.

Additionally, it was noted that many lights on the campus were surrounded by trees causing a reduction in area coverage and light intensity. A simple solution would be a regular tree trimming maintenance schedule for the campus. Table 5 illustrates a 20-year plan for the light replacement. Figures 8 through 11 provide an aerial view of the recommended locations for lighting replacement.

Life Cycle Cost Analysis

A 20-year Life Cycle Analysis (LCA) is provided based on the schedule provided in Table 5. The LCA provides options based on three types of lights, 100-watt Metal Halide, 175-watt Metal Halide, 30 watts LED lighting. The three types of lighting used in the LCA comprise lighting that is most widely used for many exterior lighting applications and would be implemented on campus to replace the current HPS lighting. An analysis of lighting types is provided in Table 6 while Table 7 provides the 20-year life cycle cost comparison

Table 4. Hall #1 and Hall #2 Data

Pole Type	Pole Height (Inches)	GPS North	GPS West	Average Proximity	Average Light Intensity	Location
STEEL	376	37° 14.231'	77° 25.242'	470.00	52.66	HALL #2
STEEL	376	37° 14.231'	77° 25.242'	352.00	18.24	HALL #2
STEEL	304	37° 14.230'	77° 25.269'	317.00	11.39	HALL #2
STEEL	225	37° 14.222'	77° 25.259'	305.40	6.52	HALL #2
STEEL	224	37° 14.215'	77° 25.253'	244.20	10.49	HALL #2
STEEL	225	37° 14.205'	77° 25.246'	297.40	5.81	HALL #2
STEEL	349	37° 14.221'	77° 25.229'	495.40	16.66	HALL #2
STEEL	354	37° 14.229'	77° 25.216'	432.40	12.06	HALL #2
STEEL	355	37° 14.229'	77° 25.216'	407.60	12.26	HALL #2
STEEL	224	37° 14.200'	77° 25.233'	288.20	4.95	HALL #2
STEEL	225	37° 14.188'	77° 25.229'	346.80	7.83	HALL #2
STEEL	135	37° 14.1847'	77° 25.221'	261.20	5.40	HALL #2
STEEL	224	37° 14.186'	77° 25.222'	294.80	9.22	HALL #2
STEEL	223	37° 14.180'	77° 25.233'	283.00	12.26	HALL #2
STEEL	221	37° 14.243'	77° 25.227'	261.00	2.86	HALL #2
STEEL	221	37° 14.170'	77° 25.236'	214.20	5.04	HALL #2
STEEL	224	37° 14.160'	77° 25.251'	209.20	13.31	HALL #2
STEEL	223	37° 14.168'	77° 25.258'	303.20	15.60	HALL #2
STEEL	223	37° 14.167'	77° 25.260'	317.80	9.57	HALL #2
STEEL	220	37° 14.164'	77° 25.2710'	290.20	6.66	HALL #2
STEEL	234	37° 14.157'	77° 25.264'	310.00	5.60	HALL #2
STEEL	224	37° 14.156'	77° 25.274'	0.00	0.00	HALL #2
STEEL	223	37° 14.157'	77° 25.278'	306.60	16.96	HALL #2
STEEL	227	37° 14.153'	77° 25.291'	333.20	4.12	HALL #2
STEEL	226	37° 14.159'	77° 25.302'	695.40	16.78	HALL #2
STEEL	223	37° 14.149'	77° 25.297'	253.60	20.51	HALL #2
STEEL	220	37° 14.155'	77° 25.304'	248.40	17.79	HALL #2
STEEL	223	37° 14.161'	77° 25.311'	237.60	9.00	HALL #2
STEEL	226	37° 14.166'	77° 25.304'	253.00	10.15	HALL #2
STEEL	228	37° 14.140'	77° 25.306'	353.20	9.47	HALL #2
STEEL	331	37° 14.147'	77° 25.313'	299.80	21.81	HALL #2
STEEL	331	37° 14.159'	77° 25.323'	252.80	26.07	HALL #2
STEEL	224	37° 14.163'	77° 25.323'	307.60	19.19	HALL #2
STEEL	230	37° 14.168'	77° 25.320'	352.40	13.42	HALL #2

continued on following page

Table 4. Continued

Pole Type	Pole Height (Inches)	GPS North	GPS West	Average Proximity	Average Light Intensity	Location
STEEL	225	37° 14.171'	77° 25.319'	310.80	5.61	HALL #2
STEEL	362	37° 14.615'	77° 25.058'	167.29	45.97	HALL #1 HALL
STEEL	356	37° 14.601'	77° 25.065'	590.86	6.44	HALL #1 HALL
STEEL	356	37° 14.609'	77° 25.067'	477.20	11.39	HALL #1 HALL
STEEL	358	37° 14.621'	77° 25.065'	558.20	13.65	HALL #1 HALL
STEEL	351	37° 14.610'	77° 25.111'	573.20	12.70	HALL #1 HALL
STEEL	355	37° 14.619'	77° 25.112'	540.20	6.14	HALL #1 HALL
STEEL	353	37° 14.613'	77° 25.135'	481.80	7.56	HALL #1 HALL
STEEL	360	37° 14.617'	77° 25.136'	456.80	9.78	HALL #1 HALL
STEEL	352	37° 14.607'	77° 25.149'	441.60	11.47	HALL #1 HALL
STEEL	330	37° 14.604'	77° 25.158'	439.60	20.89	HALL #1 HALL
STEEL	329	37° 14.590'	77° 25.144'	541.60	18.06	HALL #1 HALL
STEEL	353	37° 14.585'	77° 25.152'	385.00	11.69	HALL #1 HALL
STEEL	330	37° 14.566'	77° 25.141'	445.40	22.73	HALL #1 HALL
STEEL	351	37° 14.567'	77° 25.131'	502.40	24.36	HALL #1 HALL
STEEL	332	37° 14.547'	77° 25.136'	468.40	37.11	HALL #1 HALL

between MH and LED lighting. A quantity of 640 (80% of current campus lighting) new lighting units (pole, lamp, and accessories) was assumed for this analysis. Cost data for MH lighting was captured from the R.S. Means manual. Cost data for LED lighting was derived from average values attained from several price references. The comparisons below provide cost per year based on an assumed $0.07/kilowatt hour price for electricity. The output

Table 5. Lighting Replacement

Year	From/To
Year 1 to 5	Hall #7 to The Stadium
Year 6 to 10	Post Office to Hall #3
Year 11 to 15	Seward Hall to Hall #4
Year 16 to 20	Hall #6 to Hall #5

Figure 8. Year 1 to 5 Replacement Locations

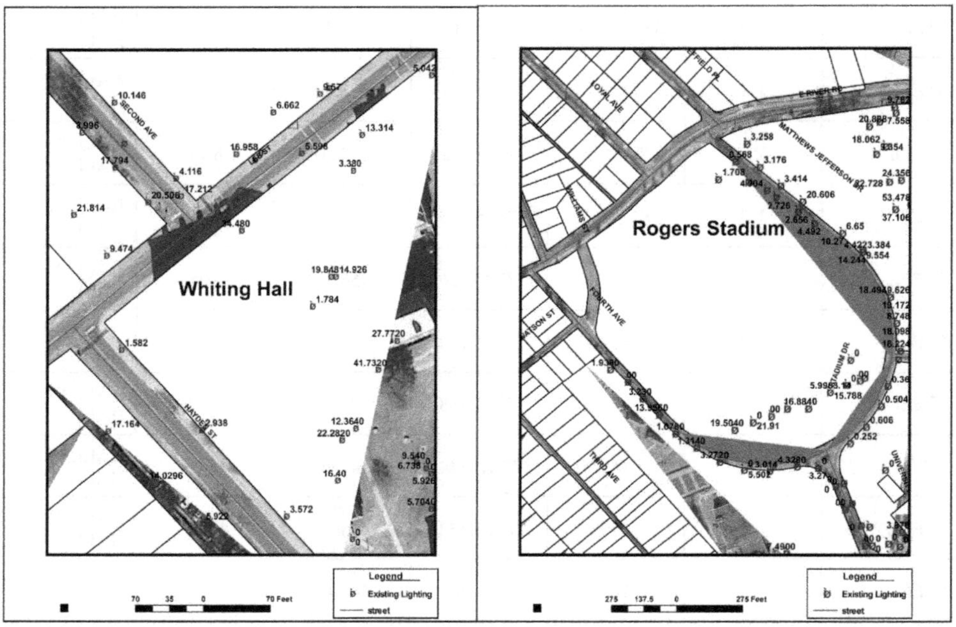

for each comparison comprises the total electrical cost per year based on the entire lifetime of each lamp as well as the yearly carbon dioxide emissions.

The results from Table 6 indicate that LED lighting provides the most efficient lighting capabilities when considering energy usage, lifetime savings, and environmental emissions. One significant disadvantage of implementing the LED lighting would be the need for additional installations (1.5 times more) to account for the low light glow thus requiring additional maintenance needs. With this occurrence, the recommended choice would be to use MH lighting. Table 7 provides the cost breakdown comparison between MH and LED lighting.

Figure 9. Year 6 to 10 Replacement Locations

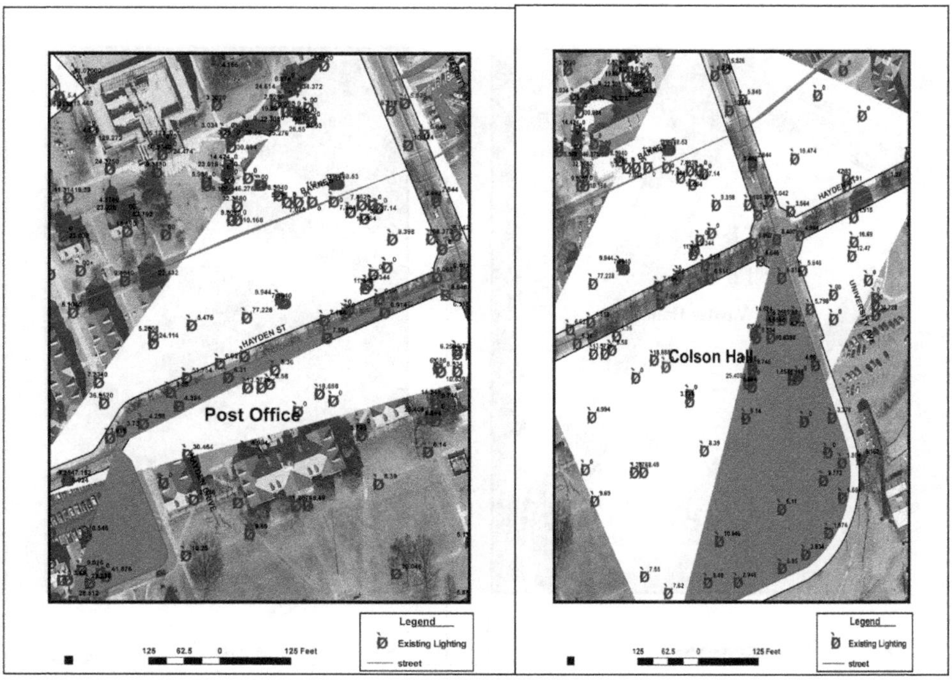

Figure 10. Year 11 to 15 Replacement Locations

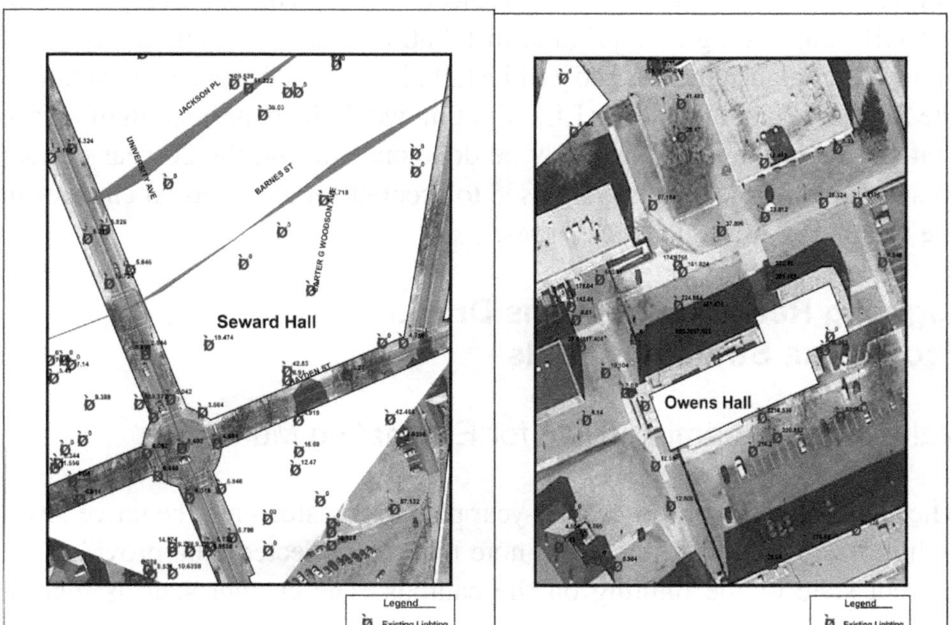

Figure 11.Year 16 to 20 Replacement Locations

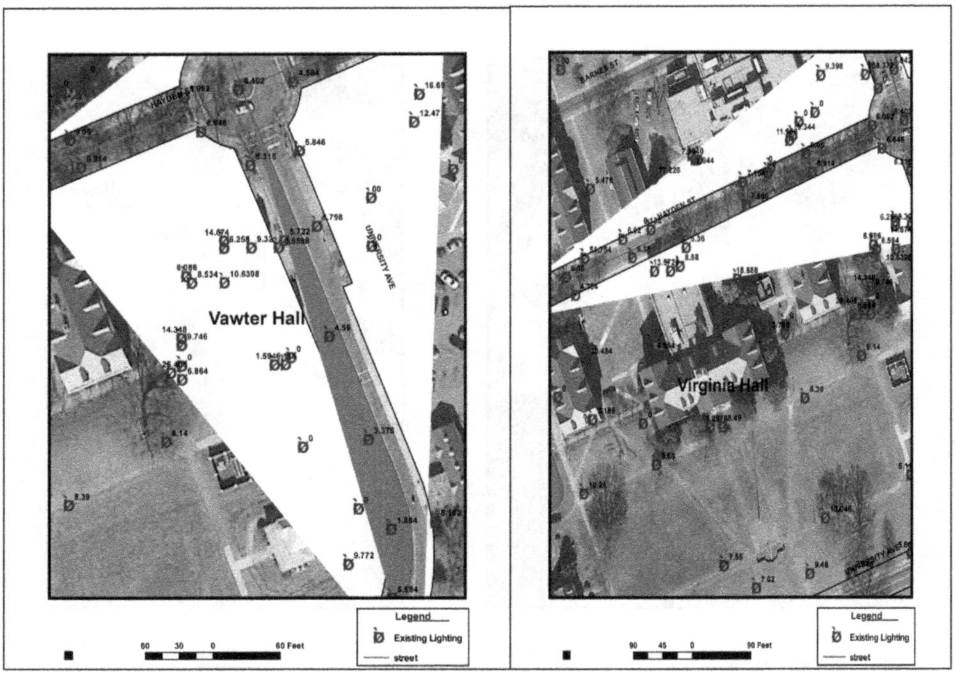

The 20-year life cycle cost indicates that MH lighting would provide for attractive returns over the life of the project. Although the rate of replacement for MH lamps is high, its original installation costs are significantly small and would balance the total cost of installation and maintenance throughout the 20 year period. Installing LED lighting would be beneficial in the sense that replacing the lamps can only be done once during the 20 year period; as such, two replacements were used to account for any unusual changes in the weather or miscellaneous issues.

Lighting Recommendations Drawn From Lean Strategies/Tools

Value Map Streaming (Plan for Eliminating Muda)

The above analysis proposes a 20-year plan of transforming the university's lighting system. It is essential to note that the collected data provided the current state of the lighting on the campus. The current state is marred

Table 6. Lighting Types Life Cycle Cost Comparison

Quantity of lights installed	640		
Price per KWH for electricity in your application	0.07		
Average number of hours of use per day	12		
Item	**MH 100 Watts**	**MH 175 Watts**	**LED 30 Watts**
Number of lights	640	640	640
Price per KWH for electricity	$0.07	$0.07	$0.07
Number of hours of use per 24 hour period	12	12	12
Current maintenance cost per light per year	14	14	5
Average lifetime (in hours) of replacement lamp	15,000	10,000	50,000
Watts consumed per lamp	100	175	30
Total current electricity cost/year	$19,062	$33,358	$5,719
Total current maintenance cost/yr	$8,806	$8,806	$3,200
Total cost per year (at current values)	$27,868	$42,164	$8,919
Average lifetime for each light (in years)	3.42	2.28	11.42
Lifetime electricity and maintenance	$318,130	$481,330	$101,810
C02 emissions (tons per year)	0.175	0.307	0.053
Comparison of savings in Electricity, Maintenance, and CO2 Emissions			
Item	**MH 100 vs MH 175**	**LED vs MH 100**	**LED vs MH 175**
Savings per year	-14,296	18,950	33,246
Savings during total lifetime of lamp	-163,200	216,320	379,520
Percentage saving	-51%	68%	79%
Reduced CO2 emissions (tons per year)	-0.13	0.12	0.25
Percentage annual reduction in CO2 emissions	-75%	70%	83%
ROI (LED/MH)			
Item	**MH 100** vs MH 175	**LED** vs MH 100	**LED** vs MH 175
Price per lamp(**single unit**)	92	600	600
Total investment Lamp Type (in bold)	58,720	$192,000	$192,000
Total savings per year	18,950	$18,950	$33,246
Total savings lifetime	64,896	$216,320	$379,520
Payback period on initial investment (years)	3.10	10.13	5.78

by energy inefficiencies, light pollution, and wastage, given the findings presented above. The future condition of the campus must be one that has energy efficiency as its priority. The tables outlined above shows the state of energy efficiency if either MH or LED lighting options are taken. The

Table 7. Lighting 20 Year Life Cycle Cost

Item	Rate	MH Individual	MH Total	Rate	LED Individual	LED Total
Original Install	1	$1,400.00	$896,000.00	1	$3,500.00	$2,240,000.00
Lamp Replacement	3	$275.25	$528,480.00	2	$300.00	$384,000.00
Ballast Replacement	2	$468.76	$600,012.80	2	$640.00	$819,200.00
Fixture Replacement	1	$726.38	$464,883.20	1	$1,000.00	$640,000.00
TOTALS		**$2,870.39**	**$2,489,376.00**		**$5,440.00**	**$4,083,200.00**

recommendation is that LED lights be installed on the campus because they offer the most energy efficiency capabilities.

Employing a VMS as shown is Figure 12 in this respect will ensure that the transformation plan eliminates energy wastage, unnecessary energy emissions, promotes sufficient lighting, cuts costs, and improves overall quality in the campus. Although the cost of LED lighting is high (mainly installation) when compared to MH and HPS lighting, the potential power savings may reduce the overall life cycle cost when compared to MH and HPS. Additionally, the cost is recouped over time and in energy cost savings. Based on a 20-year life cycle the replacement of LED lights would take place once while the MH lights would require replacement approximately six to nine times. It is essential to note that as much as the total savings would be lower in the short run, with the introduction of LED lights, especially where a similar amount is budgeted for other lighting types. The fact still remains that LED lighting is the most energy efficient, and will offer value to the campus, in the short and long run.

Total Productive Maintenance TPM and Human Resource Management HRM

Another lean tool that the university should consider adopting is TPM, with much emphasis on quality maintenance, and training and education pillars. This is especially true given the findings that trees blocked lighting area resulting in poor lighting in some areas. It was also noted that some of the poles were too long thus interfering with lighting intensity, and these poles promoted light pollution.

Figure 12. Chart showing the process of a VMS

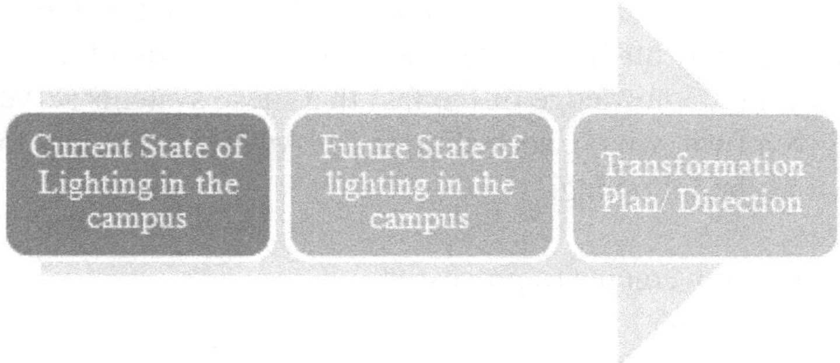

Once trimming maintenance is done on the trees next to lights would eliminate recurrence of poor lighting conditions. Secondly, conducting quality checks on the poles to make sure that they are within the range of the recommended heights would remove further wastage of light intensity and light pollution. For the university to maximize on the reduction of unnecessary costs emanating from poor lighting, it would be essential to conduct training to fill knowledge gaps that pertain to Total Productive Maintenance. In other word's, the team that is associated with campus lighting must be aligned to the need and the commitment to lean to ensure that the applicability of the above recommendations is achieved. This can be facilitated by the human resource management, where a culture of continuous improvement should prevail, among the individuals in the university that are tasked with the lighting on the campus. Such a culture would be committed to eliminating the wastes the emanate from lighting, and thus promote efficiency and sustainability throughout the 20-year period of making modifications to the Campus's lighting.

As earlier identified, a majority of the current state of lighting on the Campus is characterized by the presence of outdated lighting that employs old technology, which use outdated quality management principles of manufacturing, a scenario that appears to costs the University a lot. The study showed that the newly constructed buildings that employed modern lighting technologies had sufficient lighting. The indication from this is the fact that new buildings on the Campus like the Hall #2 which was constructed using modern lighting technologies that employed quality management systems as far as the use of advanced technology in the effort to bettering the lighting processes. Unfortunately, buildings of the caliber of Hall #2 are limited, which present the fact that the university needed to employ lean techniques to remedy its state of quality management of the lighting on the Campus.

Clearly, up to date quality management systems are vital for the Campus, especially with the knowledge of the fact that these systems serve multiple roles in a given firm, which, according to the American Society for Quality (2017) such as the following:

- Making improvements to processes
- Waste reduction
- Cutting down on costs
- Facilitating and the identification of training needs
- Staff engagement
- Setting direction for the organization

With the above insight, it is highly likely that the Campus will stand to benefit in the long run if it chooses to go with the option of using the advanced building and lighting technologies. As such, the university will be able to make an improvement of its lighting processes, as far as sufficiency and efficiency are concerned, thereby reducing wastes that emanate from these processes. In the long run, the university will be able to cut down on the unnecessary costs because of the presence of inefficient processes. The campus will also be in a position where it can identify and facilitate training needs on matters of lighting efficiency by engaging its staff members. Finally, the university will be at a position where it can set a direction of bettering processes in the organization, especially the ones that appear to be interfering with the progress of efficiency and effectiveness of the lighting on the Campus.

CONCLUSION

As earlier noted, lighting is offered for the primary aim of enhancing the comfort of the users. The aspect that ought to stand out even in the pursuit of providing high-quality lighting comfort is the fact that low consumption energy must be employed. As such, maximization of efficiency ought to be the key. The essence of efficiency was fundamental in the present study, which sought to address the lighting problems on the Campus that is in the form of outdated, and inadequate lighting. The study used lean tools to collect existing data on the light luminosity, pole locations using GPS coordinates, and other conditions. Upon the assessment and analysis of the collected data, lean solutions were proposed, which if adopted, have the potential of promoting both efficiency and high-quality lighting conditions on the Campus. The results of the study indicated that LED lighting provided the most energy-efficient option for light installation on the campus. In summary, the application of lean strategies like autonomation was seen to be fundamental to the assessment of lighting conditions, others tools like TPM, value streaming, and HRM have the potential of addressing the lighting issues, given the implementation of recommendations that were earlier outlined. The Campus stands to benefit in the form of promotion of efficiency, improvement in the lighting quality, and waste elimination, with the incorporation of lean principles.

FUTURE DIRECTION

An attempt is made to provide answers regarding the need for having sufficiency and efficiency in lighting. Some lines of research were identified as being critical to the study, and it will be in order if more studies on them are conducted to promote lighting efficiency as far as mitigating costs, reducing pollution and achieving total maintenance is concerned. The results have indicated that LED lighting, compared to the other types of lighting was the most efficient. As such, further studies should seek to demonstrate the role of the identified quality management tools in promoting efficiency, particularly when looking at LED lighting, with settings that are used being those that are looking to achieve continuous improvement in their processes. As earlier mentioned, quality management systems play certain roles that are beneficial to any organization, and so incorporating these systems into various processes is imperative. For instance, the integration of a lean tool like value map streaming in the quest to promote lighting efficiency can be used to monitor the directions towards to the future state of efficiency that the university requires. In the effort to progress forward in path of implementing quality management systems, future studies should also seek to gauge the actual role that the combination of HRM and TPM plays in staff engagement, maintenance of quality and training, especially when looking at promoting efficiency in organization, not only in lighting, but other areas as well. Overall, further studies should be geared towards confirming whether modern technologies in lighting capabilities evolve with the use up to date quality management systems, especially with the outlook for continuous improvement.

REFERENCES

American Society for Quality. (2017). *What is a Quality Management System (QMS)? -- ISO 9001 & other quality management systems.* Retrieved from http://asq.org/learn-about-quality/quality-management-system/

Andersson, R., Eriksson, H., & Torstensson, H. (2006). Similarities and differences between TQM, six sigma and lean. *The TQM Magazine*, *18*(3), 282–296. doi:10.1108/09544780610660004

Atkinson, P. (2010). Lean is a cultural issue. *Management Services*, *54*(2), 35–41.

Becker, R. M. (1998). Lean manufacturing and the Toyota production system. Encyclopedia of World Biography.

Bhamu, J., & Singh Sangwan, K. (2014). Lean manufacturing: Literature review and research issues. *International Journal of Operations & Production Management*, *34*(7), 876–940. doi:10.1108/IJOPM-08-2012-0315

Biggart, N. W., & Lutzenhiser, L. (2007). Economic sociology and the social problem of energy inefficiency. *The American Behavioral Scientist*, *50*(8), 1070–1087. doi:10.1177/0002764207299355

Boakye-Adjei, K., Thamma, R., & Kirby, E. D. (2014). *Autonomation: The Future of Manufacturing*. Academic Press.

Bureau of Energy Efficiency. (n.d.). *Energy performance assessment of lighting systems*. Retrieved July 04, 2017, from https://beeindia.gov.in/sites/default/files/4Ch10.pdf

Clarke, R. V. G. (2008). Improving street lighting to reduce crime in residential areas. US Department of Justice, Office of Community Oriented Policing Services.

Delaney, D. T., O'Hare, G. M., & Ruzzelli, A. G. (2009, November). Evaluation of energy-efficiency in lighting systems using sensor networks. In *Proceedings of the First ACM Workshop on Embedded Sensing Systems for Energy-Efficiency in Buildings* (pp. 61-66). ACM. doi:10.1145/1810279.1810293

Horm Hardware. (2017). *Lutron mini pocket lux meter lm-81lx*. Retrieved on June 6, 2017 from: https://www.horme.com.sg/product.aspx?id=5262

Javadian Kootanaee, A., Babu, K. N., & Talari, H. F. (2013). *Just-in-time manufacturing system: from introduction to implement*. Academic Press.

Krarti, M., Erickson, P. M., & Hillman, T. C. (2005). A simplified method to estimate energy savings of artificial lighting use from daylighting. *Building and Environment*, *40*(6), 747–754. doi:10.1016/j.buildenv.2004.08.007

Kurdve, M., Hanarp, P., Chen, X., Qiu, X., Zhang, Y., Stahre, J., & Laring, J. (2011, May). Use of environmental value stream mapping and environmental loss analysis in lean manufacturing work at Volvo. *Proceedings of the 4th Swedish Production Symposium (SPS11)*, 3-5.

Marquez, A. C., & Gupta, J. N. (2006). Contemporary maintenance management: Process, framework and supporting pillars. *Omega*, *34*(3), 313–326. doi:10.1016/j.omega.2004.11.003

Subramaniam, S. K., Husin, S. H., Singh, R. S. S., & Hamidon, A. H. (2009). Production monitoring system for monitoring the industrial shop floor performance. *International Journal of Systems Applications, Engineering & Development*, *3*(1), 28-35.

Torielli, R. M., Abrahams, R. A., Smillie, R. W., & Voigt, R. C. (2011). Using lean methodologies for economically and environmentally sustainable foundries. *China Foundry*, *8*(1), 74–88.

Visser, J. J. (2014). Lean in the warehouse: Measuring lean maturity and performance within a warehouse environment. Rotterdam School of Management, Erasmus University Rotterdam.

Chapter 3
A Virtual Learning Tool Design Using Lean Principles

ABSTRACT

A virtual learning tool that allows users to explore and learn the basic principle of warehousing operations using an avatar is described in this chapter. Several steps are utilized during the conception and the design phase of the application to ensure a well-built quality system. An analysis of a commonly used Lean Six Sigma principle such as the Plan-Do-Check-Act is presented to show its impacts on the conception and design of the virtual warehousing operation environments. Also, other lean principles are infused in the design of the system to ensure its user-friendliness through the use of regular tasks in an online learning environment. The use of the Plan-Do-Check-Act in the design process ensures an increased efficiency, improved quality, and better experience to potential users of the simulated warehousing operations.

INTRODUCTION

Lean is a management philosophy that focuses on the reduction of the seven wastes (overproduction, waiting time, transportation, processing, inventory, motion, and scrap) in manufacturing production and processes. By eliminating waste, quality is improved, and production time and cost are reduced. Lean process improvements occur as a result of the following basic five key steps such as studying the process by directly observing the work activities and their flow, studying the process to systematically eliminate wasteful activities

DOI: 10.4018/978-1-5225-4062-5.ch003

and their flow, coming to an agreement among those affected by the process regarding what needs to be accomplish and how, attacking and solving problems using a systematic method, and integrating the above approach throughout the organization. After applying these steps, following through with the right lean tool is essential (Bhasin and Burcher, 2006).

Lean Tools

The frequently used lean tools include Kaizen, Value Stream Process mapping, 5 S, Kanban, Error Proofing, Prevention and predictive maintenance, Set-up time reduction, reduce lot sizes, Line-balancing, Schedule leveling, standardized work, and visual management. Kaizen seeks to standardize work processes while eliminating waste using combine, simplify and eliminate as guiding words. Value Streams are actions required to create a product or service from raw material until it reaches the customer. They also seek to capture the activities taking place while people and the process are running. The 5-S creates a work environment that is clean, well organized and efficient. Kanban is a pull inventory management that improves process management by focusing on visual control of the process. Error Proofing seeks to improve a worker's ability to do their job by improving the processes. In Preventive and predictive maintenance the preventive stage maintains the equipment in good condition so that unexpected downtimes don't occur. Predictive maintenance schedules routine maintenance so that everyone knows and can plan for a machine or a piece of equipment being unavailable (Tetteh and Uzochukwu, 2014).

Setup time is the time between the productions of the last good part in one series of parts to the first good part in the next series of parts. Efforts find ways to eliminate waste in setups thus speeding up the process of setup. Reduced batch sizes allow each piece as it is created to flow from one operation to the next with no delays, storages, or work-in-process inventories. Line Balancing occurs when work is performed by each operator evenly over time with no peaks or valleys. Each worker or machine has the same amount of time, so no one person or machine is waiting for something to do or having to hurry up to keep up. Schedule Leveling establishes a program which allows the same amount to be produced daily with minimal fluctuations in demand. Standardized work comes up with a standardized operating procedure so that a job is done in the same way. Finally, Visual Management enables someone looking at a job or workspace and knows at a glance that something has been misplaced or mismanaged. With these Lean tools and techniques, a company will be able to run more efficiently getting rid of their waste. (Parry and Turner, 2006)

Six Sigma

Motorola developed the Six Sigma principles in 1980. These principles are considered to be a business strategy concept and management philosophy that sets a high standard of discipline in meeting extremely high objectives by collecting data and analyzing results to an almost zero-degree of tolerance as a way of reducing waste, defects, and irregularities in both products and services (Raisinghani, et Al, 2005). Six Sigma is a statistical concept that represents the amount of variation present relative to customer requirements or specifications. The methodologies were created by Bill Smith, who was an engineer for Motorola. He came up with this principle because of the increasing complexity of systems and products used by consumers created higher than desired system failure rates. The benefits of adopting Six Sigma are that it will enhance the ability to provide value to the customer, increase understanding of the primary business processes, will reduce waste and improve profit performance. The goal of Six Sigma is to reach 3.4 defects per million opportunities over the long term while also seeking to reduce any variability present in the process (Lucas, 2002). An improvement in just one sigma can result in a ten-fold reduction in the number of defects. At three sigma, 66,800 defects per million costing $10 per piece to fix would equal $668,000. (Barone and Franco, 2012)

Lean Six Sigma in the Design Process

The paradigms of Lean and Six Sigma are critical catalysts of change in different kinds of industries as outlined by Pepper and Spedding (2010). It is essential to point out that non-value adding activities, as determined by Gadre, Cudney, and Corns (2011), are those that a consumer would refrain from paying for. Fortunately, Lean tools seek to uncover and remove these activities, as well as to eliminate the chances of waste development to promote improvement.

Lean tools instigate operational improvement to an industry (Gadre, Cudney, & Corns, 2011). A hands-on experience is fundamental when it comes to addressing non-value adding classroom learning for example. Traditional classrooms take much longer times, and learners who need additional time to grasp a concept may be overlooked by others fast-learners (Aveta Business Institute, n.d). The best way to serve all learning styles efficiently is the introduction of simulation games in the learning process. In fact, the application of lean tools in this simulation is an icing on the cake, given that the modern industry, as evidenced by Gadre, Cudney, and Corns (2011), is pushing to be leaner by the day.

The virtual warehouse is a learning tool that allows users to explore the basic functions of warehousing operations virtually using mainly an avatar. Also, the application used a variety of 3D objects to provide feedback to users on their learning performance as they progress through the program. Another benefit to users is to be out of the constraints time and space that a physical classroom lecture environment provides (Tetteh and McWilliam, 2010). The goal is to be more appealing to the today's learners and their particular learning styles. The introduction of a virtual environment that can be used to learn the functions of warehouse operations fosters knowledge acquisition and application. However, using lean Six Sigma to aid the conception and design promotes its usability, appeal, usefulness and relevance.

Using PDCA in the Design Process

Lean Six Sigma principles were employed in the conception and design phases of the virtual warehouse through the application of the Plan-Do-Check-Act. The "Plan" is characterized by the innovative ideas infused in the application, as noted by Gidey et al. (2014). The "Do" aspect of the virtual warehouse process is characterized by the collection and documentation of the results obtained from the use of the application. The "Check" component is concerned with reviewing the errors that are identified with the application, and finally, the "Act" covers the standardization of the desired improvements to ensure that systematic learning is achieved by the users of the application (Basili, 1993). The following features of the simulated application depict the integration of PDCA in the design.

PLAN

The first contact with the application is a login screen, where the user would input his or her first and last name into the system. The login page showcases some form of organization of the various icons that can be used to allow the user to experience flexible. The lean tool, 5s, is essential in the organization of the warehouse with each component located at the right place, promoting a flexible of the system as shown in Figure 1.

Next is the *Welcome Screen* to the virtual warehouse giving access to what the application has to offer. Selecting the *About This Program* takes the

Figure 1. Initial Page

user to the next screen. Informed about the purpose of the program is then given. The *Return to the Menu* button sends the user back to the main menu with the *Explore the Warehouse* providing access to the exploration of the innovative features of the simulated warehouse operations.

The lean tool that is applicable at this point is concerned with standardized work. Standardized work, as noted by Bragança and Costa (2015) is an essential item that can be used to make standardized specification, thus establishing the best methods that ought to be employed to minimize waste. The instructional information, as well as the information on RFID, is essential in setting the pace for the procedures that ought to be taken in the exploration of the warehouse, thus minimizing the chances of having errors, on the part of the learner.

The instructions are outlined in a simple manner illustrating what is expected from the user. as illustrated in Figure 2. After selecting *Explore Warehouse*, the user receives instructions on navigation around the warehouse and different commands. The way to move between different views of the warehouse environment working knowledge of the RFID Scanner are presented. The display of a virtual image of the RFID scanner promotes the use of visual management that is an essential lean principle. The virtual images support the display of relevant information to the learner, thus promoting understanding of the functional tools of warehousing operations and environment.

Figure 2. RFID with Instructions

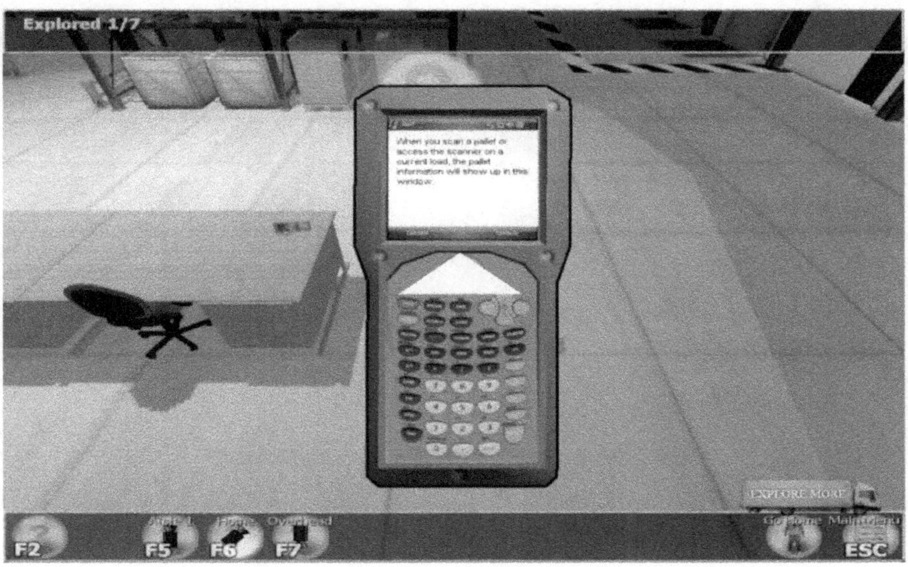

DO

The next orb teaches the learner racking systems. The concept of "Standardized work" is promoted offering the user to learn about process standardization through the simulated storage and racking areas as illustrated in Figure 3.

Users learn about the bottom, middle, and top slots of the storage racks and how items are stored based on their contents by viewing and moving around the locations and slots of the racking system using the visual tools thanks to the application of the 5s lean tool.

Another important learning feature added to the simulated system is warehouse safety. An orb provides the user with warehouse safety details such as how to be safe, and how safety must be addressed. The display of the warehouse safety instructions implies the use of a tool to standardized work. This is essential in passing firsthand information to the user about safety processes to be adhered to in the warehouse environment.

Figure 3. Simulated Racking System

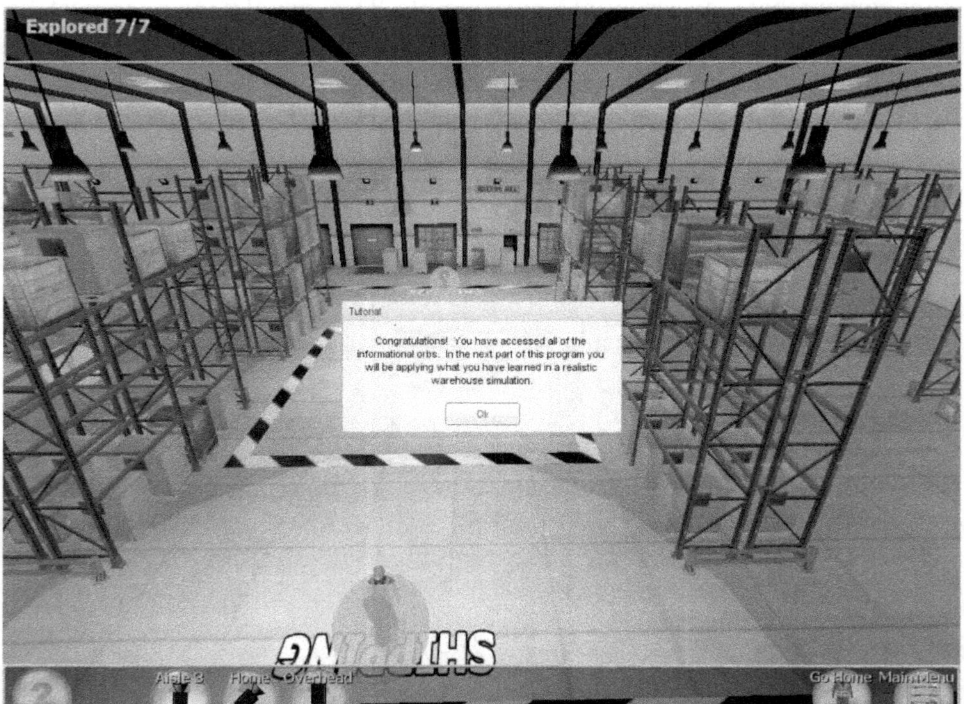

After the Warehouse safety orb, the user moves to the *Aisle Access*. The orb informs users about the aisle systems in a warehouse. The instructions regarding the usage of the aisles and their required widths are provided through still a standardized work.

Next, the user moves to the receiving area. The orb provided the user with information on the use of the area of receipt. Figure 4 shows again the use of standardized procedures, where the user learns details about the receiving area. The visual displays are also evident, indicated by the 5s floor marking tape. The safe zone demarcations on the floor reveal the 5s aspect of "setting in order," in an effort to organize the warehouse floor. The safe zone differentiation, and the writing "RECEIVING" promote a continuous flow of work in the warehouse. As such, the user grasps the need for reducing wait times in the virtual warehouse operations as shown in Figure 3.

CHECK

After learning about the shipping and staging areas, a mini screen pop up communicating to the users their access to the informational orbs and

Figure 4. Receiving Area

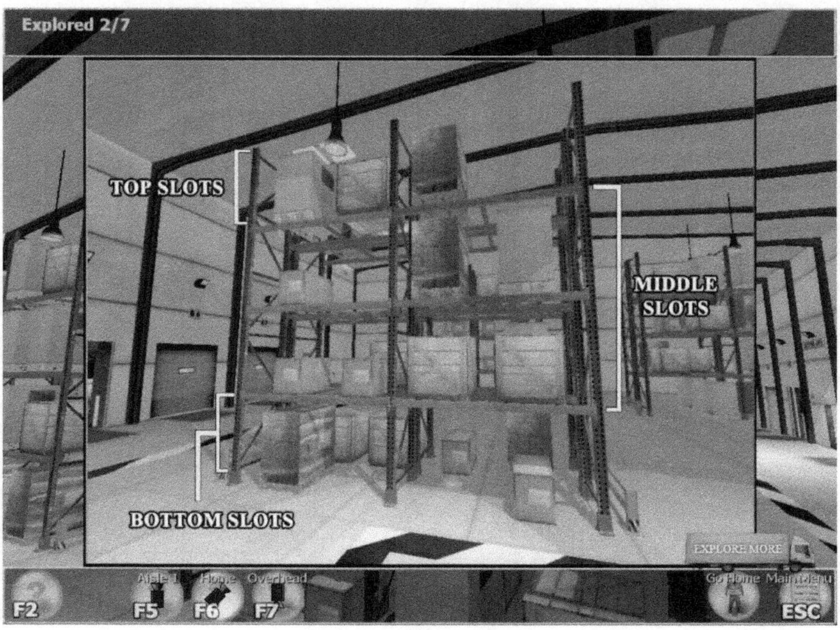

the possibility to move to the next phase of the tutorial. The presence of demarcations and the level of organization of the racks is notable in the warehouse provided further evidence of the use of 5s and visual displays as illustrated in Figure 4. For example, the clearly marked term 'shipping" allows the user to know what happens in the area, without an excessive information processing.

A screen brings the student to the next tutorial screen and explains the next part of the program. The instructions prepare the learner for navigation in a further use of the application further emphasizing the role of standardized procedures as illustrated in Figure 5.

Selecting the *Continue* icon brings the user to a reduced welcome screen of the tutorial. On the screen, Jim, the avatar is introduced and teaches the user how to operate their avatar in the first person. The use of visuals to promote a first-person understanding of instructions helps the user to be better prepared to start using the simulation. The next few steps provide information on controlling the avatar.

Figure 5. Introduction of Jim

The layout of the controls for moving the avatar forwards and backward as illustrated in Figure 6 promote an easier user's understanding of the simulation. Additionally, the organization of the warehouse is appealing, which makes learning using the simulation likely to be interesting.

Another Lean tool includes in the design of the simulation is the use of visual signals (green "Andon" indicating a go-ahead feedback), and instructions promoting the practice of the information presented to the user thus far on relevant Lean Six Sigma concepts pertaining to a warehouse management process. The user can select *Try It* to undertake the operations firsthand and have the opportunity to review his or her errors.

ACT

After moving to the green arrow visual signal, the user is presented with an RFID Scanner "how to" screen with the opportunity to walk over to a table and to pick it up and use it.

Figure 6. Simulation Controls

Figure 7. Visual Signals

As illustrated in Figure 8, a right-clicking on the mouse allow the user to pick up the RFID scanner using the avatar. The use of virtual images, Andon, and instruction guides allows the learner to understand the role of scanning in a warehouse faster and accurately. After picking up the scanner, the user is if the pickup operation is successful or not, to be allowed to move to the receiving area. Once the avatar is successfully moved to the area of receipt, the user receives the green light to scan the pallet.

The scanner displays the contents of the pallet to the user as shown in Figure 9. Through the usage of the virtual image of the scanner, the green Andon cord, and the simplified instructions, which provides feedback to the learner for proceeding to the next level, the objective is to promote a simple but clear understanding of the role of RFID in a warehouse operation environment. This is particularly the case where it comes to the determination of the contents of a pallet, in a warehouse environment to make the correct decisions. This also further the emphasis on the 5s concept.

Next, the user brings the avatar to the operating of the forklift. He or she will learn about the forklift and its operations. The instructions and image of a

Figure 8. RFID operations

Figure 9. RFID Display of Contents

forklift gives critical details to the user about the importance of transportation in a warehouses operation. The user is provided a way to move the avatar to the seat of the forklift as shown in Figure 7.

Next, the user learns the operations of the forklift by raising and lowering the forks as illustrated in Figure 10. Then, he or she proceeds to the receiving area using the forklift to pick up the pallet with the contents access using the scanner. The pallet information from the scanner helps the user to decide on the placement of the pallet on the storage racks. Also, the scanner tells the user if his/her choice is good or bad.

As depicted in Figure 11, the final operation requires the user to park the forklift back in the safe zone and press the space bar to exit the forklift and end the lesson. The description of the various features of the virtual warehouse system shows that users make use of the PDCA principles while learning to navigate in the warehouse using the forklift, RFID, avatar, exercise, instructions and other tools. However, other lean concepts are emphasised. As an example, the image of the safe zones enables the users to understand that organization

Figure 10. Forklift Operations

Figure 11.

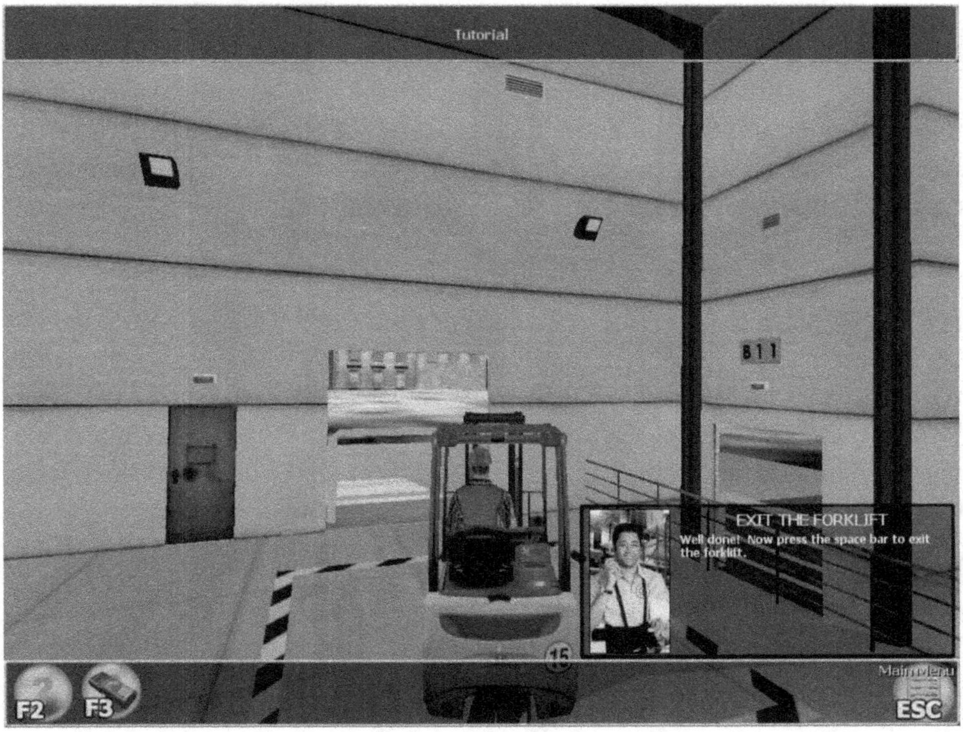

is fundamental to storage operations, particularly when the user understands the need for parking the forklift in the safe zone. Such knowledge promotes the lean tools that instigate continuous flow of work, with the elimination of stop and wait-times because of obstructions or other errors.

CONCLUSION

The simulated warehouse operations introduced in this chapter presents a brighter future of the use of Lean Six Sigma in the improvement of learning applications. It is vital to appreciate the role that simulations play in learning as far as changing how knowledge is imparted. Online simulations can be used to reinforce learning, where students can go back to the real world and apply what they have learned. This is where the future of education is headed and there is no better way to foster online learning using simulations that have been designed and improved with the application of the concept of Lean Six Sigma. Further studies should look to assess the application of Lean Six Sigma in the design of the learning simulations and attempt to see the impact that is obtained at the various levels of learning particularly when looking at how the users are able to apply what is learned in real life. In summary, further studies should move beyond the use of lean Six Sigma to design and modify online simulations, but look to see how beneficial the LSS strategies are when paying attention to how users become knowledgeable on a particular concept such as warehouse operations.

REFERENCES

Aveta Business Institute. (n.d.). *The Online Vs. Classroom Based Six Sigma Training Debate*. Retrieved July 05, 2017, from http://www.sixsigmaonline.org/six-sigma-training-certification-information/the-online-vs-classroom-based-six-sigma-training-debate/

Barone, S., & Franco, E. L. (2012). Six Sigma methodology. *Statistical and Managerial Techniques for Six Sigma Methodology: Theory and Application*, 1-21.

Basili, V. R. (1993, September). The experience factory and its relationship to other improvement paradigms. In *European Software Engineering Conference* (pp. 68-83). Springer. doi:10.1007/3-540-57209-0_6

Bhasin, S., & Burcher, P. (2006). Lean viewed as a philosophy. *Journal of Manufacturing Technology Management*, *17*(1), 56–72. doi:10.1108/17410380610639506

Bragança, S., & Costa, E. (2015). *An application of the lean production tool standard work*. Academic Press.

Gadre, A., Cudney, E., & Corns, S. (2011). Model development of a virtual learning environment to enhance lean education. *Procedia Computer Science*, *6*, 100–105. doi:10.1016/j.procs.2011.08.020

Gidey, E., Jilcha, K., Beshah, B., & Kitaw, D. (2014). The plan-do-check-act cycle of value addition. *Industrial Engineering & Management*, *3*(124), 2169–0316.

Lucas, J. M. (2002). The essential six sigma. *Quality Progress*, *35*(1), 27.

Parry, G. C., & Turner, C. E. (2006). Application of lean visual process management tools. *Production Planning and Control*, *17*(1), 77–86. doi:10.1080/09537280500414991

Pepper, M. P., & Spedding, T. A. (2010). The evolution of lean Six Sigma. *International Journal of Quality & Reliability Management*, *27*(2), 138–155. doi:10.1108/02656711011014276

Raisinghani, M. S., Ette, H., Pierce, R., Cannon, G., & Daripaly, P. (2005). Six Sigma: Concepts, tools, and applications. *Industrial Management & Data Systems*, *105*(4), 491–505. doi:10.1108/02635570510592389

Tetteh, E., & McWilliams, D. (2010). Simulation design for off-line training of practical lean manufacturing concepts for visual inspection. *Simulation & Gaming*, *41*(4), 553–567. doi:10.1177/1046878109334334

Tetteh, E. G., & Uzochukwu, B. M. (Eds.). (2014). *Lean Six Sigma Approaches in Manufacturing, Services, and Production*. IGI Global.

Chapter 4
Reducing Hospital Costs by Reducing Supply Waste

ABSTRACT

Hospitals lose a large amount of money through waste as defined by lean. Processes have been in place many years at medical facilities and waste is often accepted as unavoidable in the rightful "patient first" mentality. The Plan, Do, Check, Act cycle is used for the control and continual improvement of the waste management process at the hospital where the study took place. By adapting and utilizing lean and statistical tools such as scatter diagrams, histograms, Pareto charts, fishbone diagrams, and control charts, waste can be identified and reduced or eliminated. The use of the statistical tools provides a visual presentation of the message that the team conveyed. They are especially useful to summarize the information that would normally require much time and effort to elucidate its complexity. Further, they provide a visual summation with still a great deal of details leading to increased satisfaction, efficiency, and creates competitive advantage.

INTRODUCTION

Supplies in hospitals account for a large part of the expense of a hospital's budget. If supplies are not carefully monitored, costly waste can quickly accumulate. The redesigned process for controlling supplies will be implemented throughout the hospital after gaining stakeholder buy in and demonstration of successful results on two nursing units and thereby reducing costs for the hospital as a whole.

DOI: 10.4018/978-1-5225-4062-5.ch004

The major steps will utilize the Plan, Do, Check, Act cycle. To further utilize the scientific method, Identify and Define have been added preceding Plan, Do, Check, Act.

1. **Identify:** Identifying and understanding why supply waste is a problem worthy of being addressed by a project by early research in a short and concise statement. A Gantt chart will also be developed. (IET Option(s) IET419, Total Quality Improvement, Week 1)
2. **Define:** In this phase the problem will be clearly defined and the goal will be stated. A team will be built and stakeholders identified. (IET Option(s): IET319, Quality Control, Weeks 2 - 3)
3. **Plan:** Data will be gathered and the current process will be analyzed using techniques and tools of Quality Management. (IET Option(s): IET317, Lean Systems, IET319, Quality Control, IET419, Total Quality Improvement, Weeks 4 - 5)
4. **Do:** Process changes or physical layout changes resulting from the Plan phase will be implemented in the test nursing units. (IET Option(s): IET317, Lean Systems, IET419, Total Quality Improvement, Weeks 6 - 7)
5. **Check:** Time studies and economic analysis will be compared before and after the process change. If goals are not met reassessment of changes will occur. (IET Option(s): IET 317, Lean Systems, IET319, Quality Control, Weeks 8 - 9)
6. **Act:** If the results are positive the new process and changes will begin to be implemented throughout the hospital. (IET Option(s): IET319, Quality Control, IET419, Total Quality Improvement, Week 10)

Identify

The primary goal of this is to modify or design a process to reduce supply waste by increasing compliance of nurses charging supplies out to patients when used. Additionally, reducing calls made to the materials management department and decreasing the occurrence of temporary supply shortages within the inpatient supply storage areas are secondary goals.

Preliminary data will be gathered from each inpatient nursing unit will be evaluated for the amount of "lost chargeables" it produces. In this stage, data of PAR compliance vs. lost chargeables will be collected from the PAR system for each inpatient unit to be studied. Before the problem is properly identified it must be determined that the right issue is being examined. The simplest way to determine if a cause-and-effect relationship exists between variables is to plot a scatter diagram (Besterfield p. 50). Once the issue, stakeholders, scope and desired results are identified, this data will be utilized as a guide for the rest of the steps.

Define

Once the issue has been identified by the continuous improvement team, a larger team consisting of the stakeholders affected will be created. The identified problem will be discussed among the stakeholders to solicit input. PAR Excellence software and hardware is in use and determines PAR compliance (how much in percent of lost chargeables are occurring) automatically on each nursing unit. The metrics of PAR compliance percentages and reduced lost chargeable cost, will be the primary metrics for which the success of this project will be measured. Next the goals concerning the amount of PAR compliance percentage and reduced cost will be discussed, agreed upon and stated.

Finally, a histogram will be constructed for the nursing units PAR compliance to help visually determine the poorer performing units. PAR compliance percent's will be broken down weekly as the histogram data. After the problem is clearly defined, goals stated and agreed upon then a project charter will be created.

Plan

A Pareto analysis will be performed on items located within the supply room. The software is able to determine which and in what quantity the individual supplies are not complied with. Determining why those items are the worst in compliance will be a goal as increasing their compliance will raise PAR percentage the most dramatically.

A flowchart will be created that will demonstrate the supply chain for the nursing units selected for improvement. Throughout the plan portion of PDCA, an Ishakawa (fishbone) diagram will be constructed that will help determine the candidate causes of poor PAR compliance performance (Aikens p. 199). The "5 why" tool will also be utilized in conjunction with the fishbone diagram. The physical layout of the supply room will be evaluated for size and ease of use vs. other more successful inpatient nurse units. Based on the results of the physical evaluation and a spaghetti diagram, 5S will be utilized. Time studies will be performed before any changes are made to the physical layout so that a before comparison can later be evaluated. After all data is analyzed, a report is created for management with recommendations and justifications.

Do

After the planning phase the team will reconvene and discuss the findings from analyzed data. Once every part of the process is understood, the organization can seek to eliminate steps, combine steps, rearrange steps or simply them (Meyers, 2002 p. 66). After management approval is obtained, the recommendations will be implemented. The physical changes to the layout will be made as well as any changes to the supply chain itself. Other opportunities for improvement such as staff education or hardware/software changes to PAR that were found during the plan portion will also be addressed.

Check

After implementation has occurred, ample time will be needed to see the impacts of the changes that occurred. Data will be collected before, during, and after the Do stage from the PAR system to see if compliance has increased and the cost of lost chargeables has decreased. Interviews will be gathered to gauge the level of continuing employee buy in. During this phase a control chart will be constructed of the past PAR compliance weekly figures and each week the new numbers will be added to the run chart. A new time study will be performed to compare before and after results.

Act

In this final stage, an economic analysis based on findings will be present in the report as well as the most impactful steps taken. Upon management approval a Gantt chart will be made and the plan replicated through the rest of the nursing units. As this is continuous improvement, while other nursing stations are beginning to adopt the changes that resulted from the original two nurse stations, a new Pareto analysis will be performed of supply items in the original units to continue to work on the next supply issue that have the lowest PAR compliance.

Identify

The process engineering team at the local hospital is concerned about the amount of lost chargeables that that have occurred because of consistently low PAR scores. It has been further explained that the PAR Excellence system is a "pull" system that utilizes a hardware wand and buttons placed near supplies that is used to charge a supply and quantity out to a specific patient. When it is not used properly Material Management cannot in turn stock the patient supply rooms used by the nursing floors efficiently. She suspects that the lost chargeable cost rises when PAR compliance falls. The task for the process engineers was to look at the current state process to determine root cause and aid in improving the compliance percentages.

Management has approved the project for process engineering and the Identify, Define, Plan, Do, Check, Act method will be used for process improvement. To determine if Sheila's hunch is correct, data in the form of monthly PAR compliance percentages and corresponding lost chargeables are

Figure 1. Scatter diagram displaying negative correlation of PAR Compliance vs. Lost Chargeables

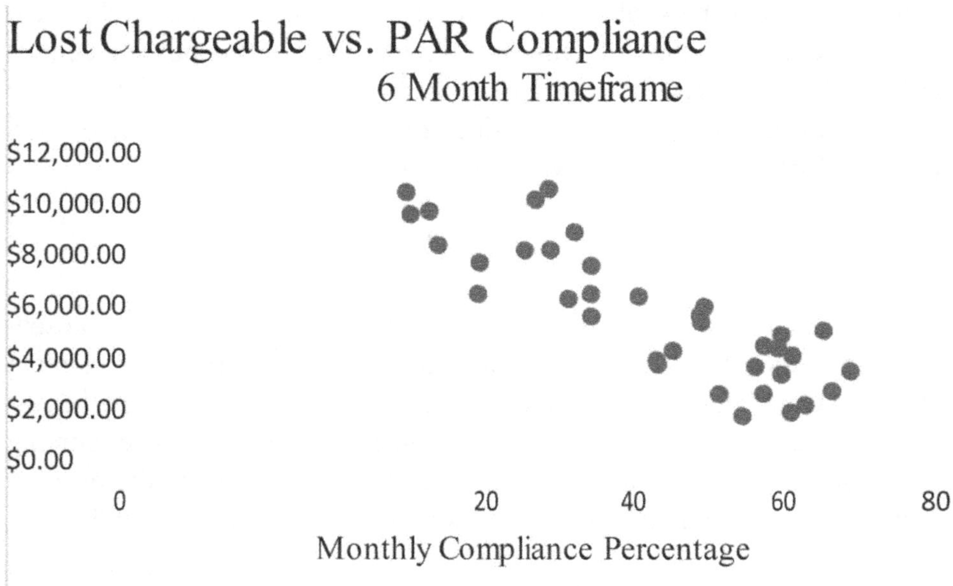

gathered from the PAR Excellence system and a scatter diagram is created to see if a correlation has occurred. The scatter diagram is charting data for six floors, for six months (Figure 1).

The scatter diagram does show a strong negative correlation between cost rising (variable 1) and monthly PAR compliance falling (variable 2). With solid data supporting any assumptions, stakeholders are identified. The overall objective is to reduce lost chargeables by increasing PAR compliance and the scope expanded upon in the next phase, Define. First, the problem is clearly identified.

Identify Statement: Low compliance with the PAR system by the nursing staff is causing higher lost chargeable supply cost for the hospital.

Define

Now that the problem and team members have been identified, further data will be used to determine which floor has a high frequency of low par compliance scores. Using 15 months of historical data that has been gather during the Define step, the units can be compared visually by using a simple but useful tool, the histogram. The Bins represent percentage. Thus, the higher

Table 1. Historical monthly average data gathered from six nursing units

6th Floor Unit 1	6th Floor Unit 2	7th Floor Unit 1	7th Floor Unit 2	8th Floor Unit 1	8th Floor Unit 2
70.9	67.3	33.1	21	61.1	64.8
68.4	45.1	36.1	21.2	58.1	62.9
63.1	42.5	30.7	12	51.3	59.2
59.3	36.2	28.7	14.5	50.7	53.3
61.6	33.9	27.3	15.7	47.1	56.4
61.6	36.2	30.5	11.4	50.9	44.9
69.4	45.6	56.4	35.5	58.1	53.9
64.3	65.4	85.3	65.5	71.3	59.6
58.6	60.5	58.6	58.8	60.8	58.6
48.5	50.7	48.5	51.3	67.7	51.3
44.4	51	59.6	27.9	70.8	62.5
51.8	51.2	55.6	16.4	74.9	75.5
56.4	58	45.6	33.6	70.6	67.7
63.4	43.2	48.9	31.2	62.4	61
60.7	40.5	37.9	27.2	55.7	59.8

frequency that occurs on the left of the charts with the lower percentages, the bigger the compliance problem for that unit.

Historical gathered date for each floor, spanning 15 months (Table 1).

Histogram for all units (Figure 2).

From the raw data:

Average = 50.65 | Median = 54.75 | Mode = 58.60

The Average < Median < Mode and we can determine that the histogram is skewing left.

By visually analyzing the histograms it can be determined that both 7th floor units are performing worse than the others. It can be seen that 7th floor Unit 1, only has one occurrence of an 80-89, and it is separated, so it is likely an outlier. 7th floor unit 2 also had only one occurrence of a PAR compliance percentage in the 60-69 range. Using this evidence, the team decides to focus

Figure 2. Histogram showing more activity skewing toward lower compliance

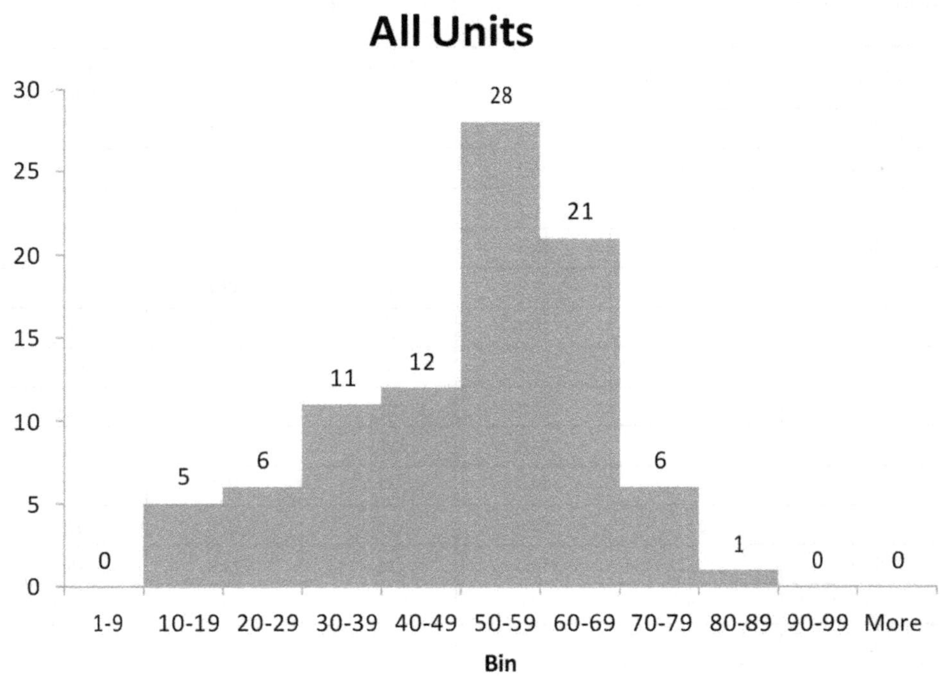

Figure 3. Histogram comparison of all six units. Notice 7th floor

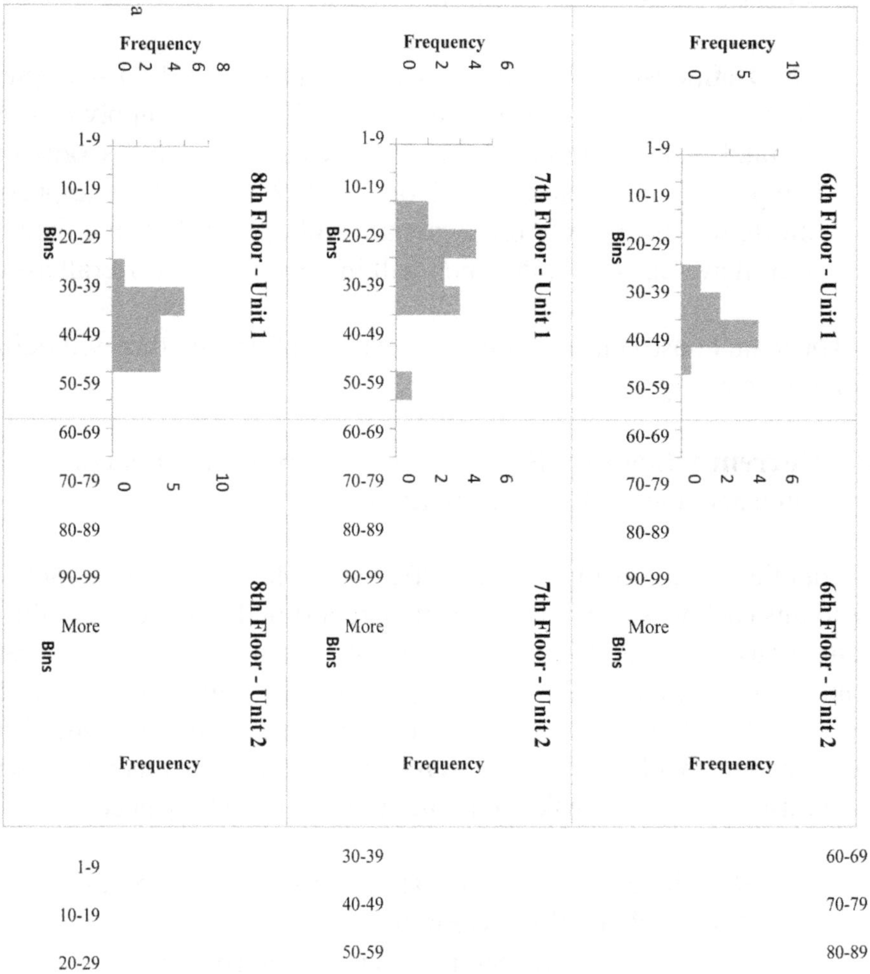

Figure 4.3: Histogram comparison of all six units. Notice 7th

on these two areas as the target for improvement, Patsy Sexton the manager of this department is added to the stakeholder list.

In the design phase the identify statement is expanded upon and the scope becomes clearer. Based on the raw data 7th floor – Unit 1 has an average PAR compliance of 45.5 and 7th floor – Unit 2 has a PAR Compliance of 29.5. The goal as decided by the team is to raise these areas to the current combined average of 50.65.

When working to find a root cause problem it is important to continually narrow down and be concise. It has been determined that there is a problem, and that the worse location is the 7th floor. Furthermore, we see that 7th floor - Unit 2 is the worst performing floor.

- **The Define Statement is:** Low compliance with the PAR system by the nursing staff is causing higher lost chargeable supply cost for the hospital. The 7th floor has been shown to have the lowest scores for PAR compliance, particularly 7th floor – Unit 2. Continuous Improvement efforts will begin on these two units and attempt to bring them to the overall average of 50.65 which will in turn raise that overall average.

The team constructs a charter during the Define step that will define the problem and scope.

- **Current Situation:** (Background on the problem or condition; how it came about and what is happening)

Supplies are not being utilized efficiently. Materials management has to do counts each weekday and sends out a report on lost chargeable that show many thousands of dollars as lost chargeable items. Nursed do not PAR out items as they should. Materials brings too many items and places them in shelves, where they sometimes expire. Old items are not brought FIFO and are pushed further back in bins. There are old items/bins that nurses say they do not use often and it makes it harder to find items they need.

- **Key Stakeholders:** (identify team members and support personnel)
  - **Billy Tackett:** Process Engineering
  - **Sheila Ramey:** Director, Materials Management.
  - **Kimberly Johnson:** PAR Coordinator
  - **Debbie Parsons:** CNO/COO
  - **Patsy Sexton:** Manager of 7th floor
- **Importance:** (Why a problem)

Table 2.

CONTINUOUS IMPROVMENT CHARTER **Date:** March 2016 **Department/Section:** 7TH Floor Unit 1 and 2 **Lean Target Area to Be Improved:** Supply Rooms **Manager:**_______**Supervisor:**

Inefficient use, hospital could be missing revenue or could save money by stocking the correct amounts instead of overstocking. Additionally, many calls are made to materials management for on demand deliveries when stock runs out interrupting their day and causing wait.

- **Waste Identified:** Transportation, Overproduction, Inventory, Waiting
- **Scope/Boundaries:** (Beginning and ending of process/project)

Research and collect data of supply usage, method and processes of using supplies, and restructuring supply rooms.

- **Measurements:** (i.e., Cycle time, WIP, inventory, setup, 5S, etc.)

PAR Compliance, Amount of Money Lost, Amount of Money saved, Amount of calls made for emergency deliveries. Primarily PAR Compliance measurements.

Planning

The planning phase is a major step in PDCA. It is during these two weeks that the process is walked, time studies performed, data gathered, the layout recorded, and root-cause analysis performed. It is important to continually narrow what may be causing the problem. Data has already been used to determine if there was a problem, and where the problem is. Further data is available to determine what specific items have the lowest compliance before visiting the floor. A report is generated with the quantity of items not complied with on 7th floor

- Unit 2 and a Pareto analysis performed. The Pareto is a tool used in to identify the most frequently occurring problems and what cumulative percentage they represent (Besterfield pg. 19).

Table 3.

Performance Measure	Current Measurement	Goal Measurement
7th floor Unit 1	45.5	50.65
7th floor Unit 2	25	50.65

Table 4. Two months of data gathered from 7th floor – unit 2 showing the amount of times and item came up deficient with PAR compliance

Item	Number Missed	Cumulative	Cumulative %
ST ADMN PRIMARY PUMP W/CART15GG 12538-28	872	872	21.74
PLUG IV MALE ADPTR LL LIFESHIELD	537	1,409	35.13
ST IV SECONDARY IVPB 15 GTT 32IN 2567-50	499	1,908	47.57
KIT IV START W/STATLOCK STER	495	2,403	59.91
CATH IV INSYTE 22X1 BLUE 381423	477	2,880	71.80
CATH IV INSYTE 20X1.16 PNK 381434	323	3,203	79.86
CATH IV SHIELD SAF-T-INTIMA 22GAX0.75IN	288	3,491	87.04
CUFF BP 2 TB SCRW LNG SOFT-CUF ADLT	281	3,772	94.04
CNSTR SUCT SHUT OFF FLTR 1200ML 71-3002	151	3,923	97.81
ST EXTN REGULATOR DIAL-A-FLO 3ML 18IN	88	4,011	100

Figure 4. Two months of data for 7th Floor – Unit. Three items are causing 50% of issues.

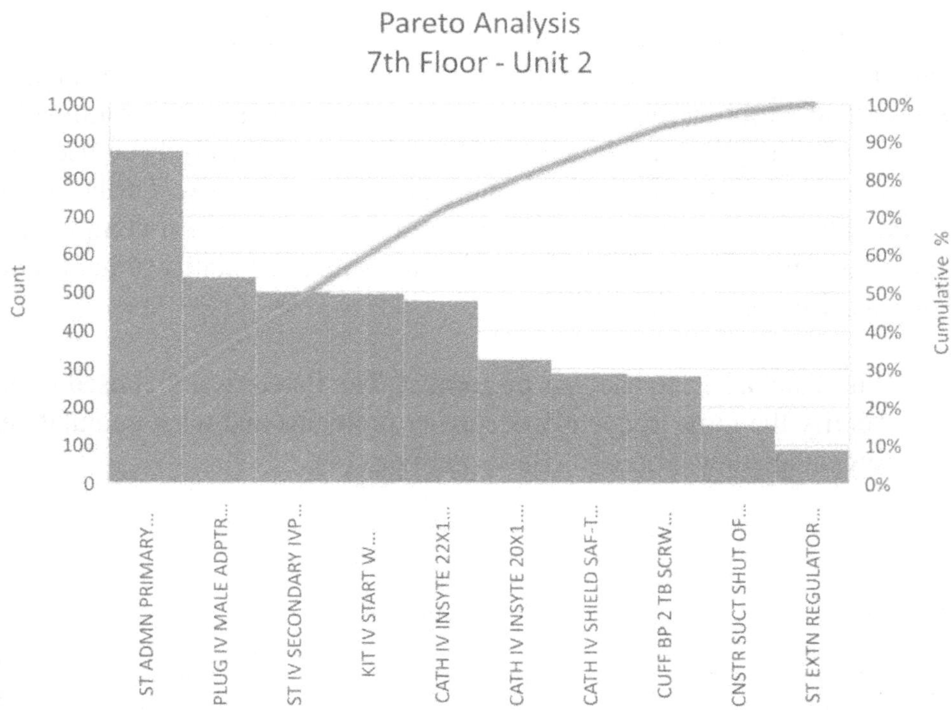

Figure 5. Current state shows many areas for waste

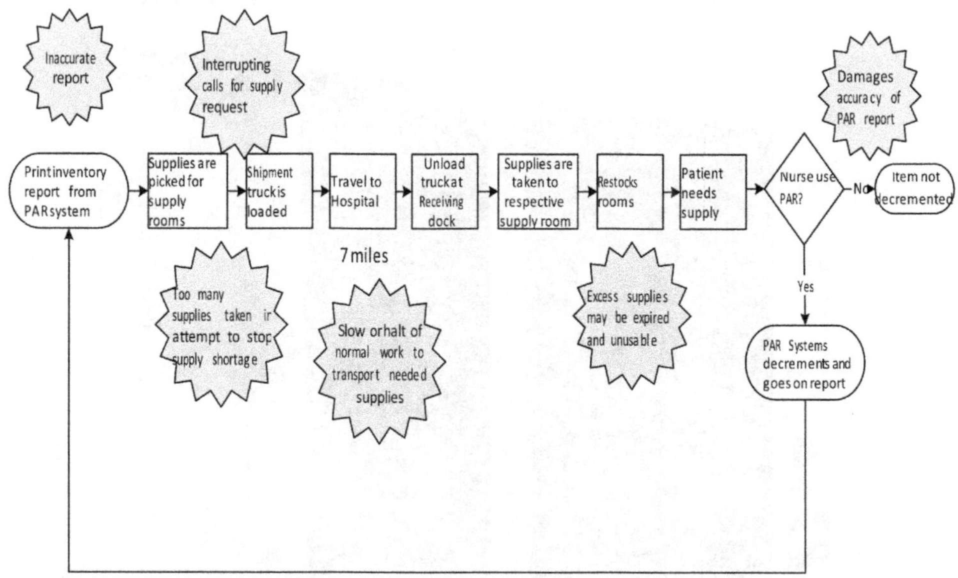

After determining the top three items associated with the compliance issue by way of the Pareto chart, the site (Gemba) was visited and the supply chain mapped below.

Observations of both the supply chain and the nursing staff occur over the next two weeks. A flowchart was created of the current state as the supply chain is observed. By PAR compliance being low many types of waste are created upstream.

Photos were taken of the supply room before in order to make a comparison during the Plan Phase (Figures 6-9).

5-Why's

The 5 why method is a tool that helps us determine root cause by asking why until we drill down to the root. Note that the "5" is an arbitrary number, it can take more or less to get to the root cause. Referring to the Pareto Diagram in Figure 4 we can see that the ST ADMIN Primary Pump is about 20% of the problem and the 5 why method is employed.

1. Why is this item the worst? It's out of reach of most staff.
2. Why is it out of reach? Lack of space and it had to go somewhere.

Figure 6. The hardware PAR system. Nurses use the wand on the left to press buttons on the supplies then touch the button of the listed patient on the board.

3. Why is there a lack of space? Too many items in the room.
4. Why are there too many items in the room? Staff request items to be stocked but when they are no longer needed do not request them to be removed.
5. Why don't staff request no longer used items to be removed? No designated person to take charge of supply room.
6. Why is there no one taking charge? No one ever has. That's the way it has always been done.

This 5 Why's happens to pertain to most of the items found in this supply room and you can certainly drill down further by asking, why is there no designated person. 5 Why examines one reason at a time and drills down to its root. Other 5 whys might answer why the unit as a whole has the worst PAR compliance but would ultimately reach the same conclusion. Too many items make it overcrowded and hard to find items and no one is taking charge.

An Ishikawa diagram also helps us to find root cause visually. It is constructed with the observation in the supply room. Notice the Environment

Figure 7. The supply room is small and this large trash can restrict access to the wand. Since nurses deal with direct patient care, they do not like moving the trashcan to access wand.

section in Figure 10 and compare it to the 5 why above. It differs from 5 Why in showing multiple issues that can lead to the root cause in contrast to 5 why which is specific to one issue at a time.

In order to change the color coded duct tape to something even more visual, easily labeled and containing more supplies a cost estimate was completed to find bins that could be hung along the wall. It is decided during planning that the supplies can be divided into six different categories. Miscellaneous supplies (gray/white), wound care supplies (green), heart and blooded related supplies (red), respiratory supplies (blue), personal hygiene and cleaning supplies (purple) and urology supplies (yellow).

The last two steps used in the Plan phase for this project is a time study and a spaghetti diagram. A simple stopwatch time study analysis is performed each day that shows the average time to find an item for a staff member that

Figure 8. The supply room current layout. Color coding was implemented from the beginning but could be improved. Items commonly used together are not grouped next to each other.

**For a more accurate representation see the electronic version.*

Figure 9. A large refrigerator sits in that supply room and is only filled up halfway. Items were being stacked on top of it due to lack of room for storage.

Figure 10. An Ishikawa (Fishbone) diagram visually showing problems on each topic

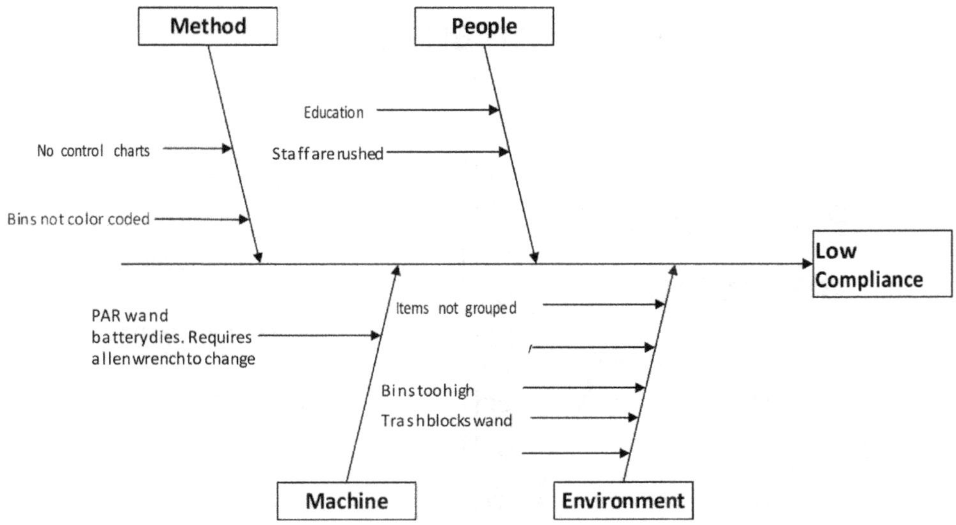

normally works on this unit to be about 30 seconds. For staff member who float and cover for this floor that raises to 1 minute and 20 seconds.

Finally, the spaghetti diagram is used to show the movement of staff as they search for items (Figure 11).

In the planning stage we have found that supplies are out of reach, too many supplies that are unused or have very low use are stocked making it harder to find items needed. There is no real 5S in place, objects are in the way, space is underutilized and motion waste is occurring.

There is also no daily, weekly, or monthly tracking of the software numbers. No one is a champion for the area and keeping it updated because "that's the way they've always done it."

The team meets back together and discuss plans on running reports to see what items have low or no usage and removing those, reducing unused inventory and decreasing search time for items. The team also discusses placing colored bins in to help visually determine where items go and grouping them together should also decrease search time. After research of bins and supplier is conducted a decision is made to purchase the Quantum Magnum hanging bins found online: http://www.cartsbinsmats.com/store/c42/QMS_QUANTUM_MAGNUM_SERIES_BINS.html at a cost of $3395.04.

A visual aid such as colored masking tape will be used to limit made for how high the new colored bins and the laminated legend are to be placed. The electronic PAR board may be moved to make movement flow better.

Figure 11. Current movement of the staff obtaining supplies

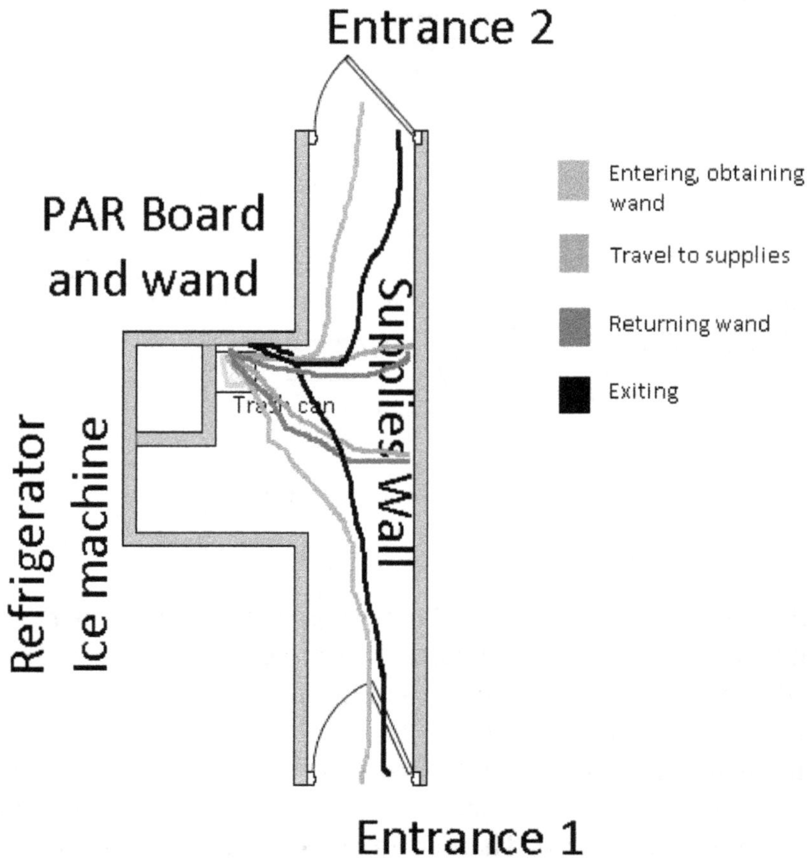

**For a more accurate representation see the electronic version.*

Also it is agreed upon that the PAR wand batteries are to be checked every two days. The refrigerator will be taken out since it's only half used to make room for more supplies. A new place for the trash can or a smaller one will be found. Finally, the director of this nursing station and supply room will also name a champion for the area.

Do

Do begins with an outline to show the procedures to implementation, this is often the most expensive step because it creates changes in environments and processes.

1. Create recommendations for the Do step from information gathered during planning.
2. Take recommendations to the process team for approval and ensure everyone is on the same page.
3. Recommendations are then given to management for approval.
4. In the event management approves some or all of the recommendations, begin implementation, which will be different for each change made.

During this phase of IDPDCA the recommendations are made and implemented from the planning stage. These are the recommendations that the team believes will solve or help to alleviate the problems so it is important to implement it correctly.

- Relocate the garbage from the room.
- Relocate the PAR wand and board closer to the entrance.
- Remove items that are not critical that have not been used within the past 3 months.
- With removed items freeing up space, move the top row of bins down to decrease height of supplies.
- Remove refrigerator and replace with closable cabinet.
- Change the cage bins with colored duct tape to colored plastic bins.
- Group together items that are often used together.
- Have materials management replace & check the PAR wand battery and log that it has been checked one per week.
- Clean and shine the supply room.
- Educate nurses as to the importance of using supplies correctly.

After the continuous improvement team agreed on the changes, it is then taken to management to review. In this project these changes had low cost and all were approved.

The garbage bin has been moved to a different room. Colored bins have moved in, replacing the cage versions and also lowered to make for easier reach. Reports of the PAR system are utilized to see which items were no longer used and during the switch out to colored bins they were tagged as part of 5S and removed. Items that nurses often gathered together were place next to each other to cut back on time searching. The area is also shined and better lighting has been installed. Pictures were taken after these recommendations were implemented (Figure 12 and Figure 13).

Figure 12. The refrigerator has been removed and a cabinet to hold more needed supplies is installed

Check

After the implementation of the Do phase of IDPDCA, check takes place. During this phase the new process is checked to see if recommendations implemented have the desired of effect to improve compliance. One way of doing this is by creating using a statistical process control tool called a control chart. In this case, 4 samples are taken per month.

Since the sample size less then than 10, an Average – Range (X_{bar} – R) chart is used check the process. The double x bar, process standard deviation, and sample standard deviation are calculated along with the upper control and lower control limits. These calculate respectively to 30.63, 12.74, 6.37, 11.5, and 49.7. The large standard deviation shows that this process is experiencing a large spread of variation. With these calculations the X_{bar} chart is constructed as seen in Figure 14.

Figure 13. Cages holding supply bins have been replaced with colored bins. Red masking tape visually limits height

**For a more accurate representation see the electronic version.*

Table 5. Data gathered 4 times per month

	Week 1	Week 2	Week 3	Week 4	X-Bar
November	15	17	16	22	17.5
December	22	37	35	30	31
January	18	17	16	25	19
February	24	36	35	25	30
March	34	32	31	35	33
April	50	55	48	60	53.25

In the X_{bar} - R chart presented, the tracked grand average line ($X_{bar\text{-}bar}$) shows a large jump in compliance. It is reflecting data before, during and after the change in processes from the Do step of IDPDCA. Although this shows that the process has an out of control point, it is to be expected because the recommendations should raise the compliance. So this is an assignable cause that is explained. After more recording over the next few weeks the

Figure 14. Xbar chart showing the variation in the process. In this case the 53.25 is desired as it demonstrated higher compliance as implementations are recommended.

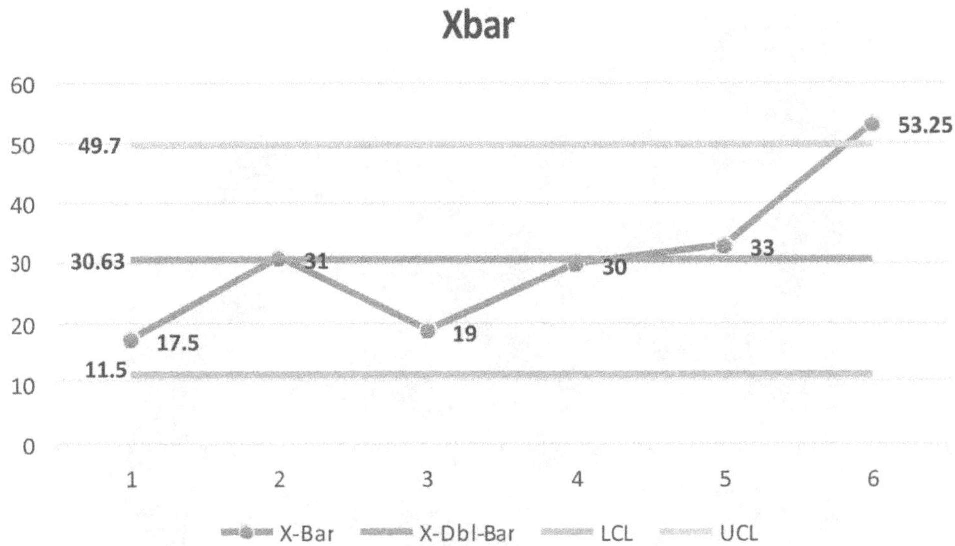

**For a more accurate representation see the electronic version.*

X_{bar} - R chart will be revised to calculate the new normal which will be the final chart for the nursing unit to utilize

Before it can be determined that the X_{bar} chart is accurate, the range (R-Chart) must be assessed first. Using the same data. The R_{bar}, Double X-bar, LCL(x_{bar}), UCL(x_{bar}), LCL(R), and UCL(R) are calculated respectively to 9.8, 30.63, 23.46, 37.79, 0, and 22.44. The R-Chart is shown in Figure 15.

The R-bar chart is constructed and is assessed. The data is close to the center line and there are no out of control points on the range chart thus the X_{bar} chart can be said to be reliable.

Additionally, an after time study is completed with the results of the time studies showing a new average of 14 seconds for nurses working the area and a drop to 20 seconds for floating nurses.

Act

In the final phase of the IDPDCA, it is decided by staff and management to pursue this project with the other supply stations. Calls have been reduced to materials management and as the Xbar R chart showed, compliance is beginning to rise. Below is the process map for the final process (Figure 16). Notice the reduction in interruptions and errors of inventory.

Figure 15. The R chart shows the R line hugging near the center line and nothing out of the ordinary thus demonstration the Xbar chart in figure 14 is reliable

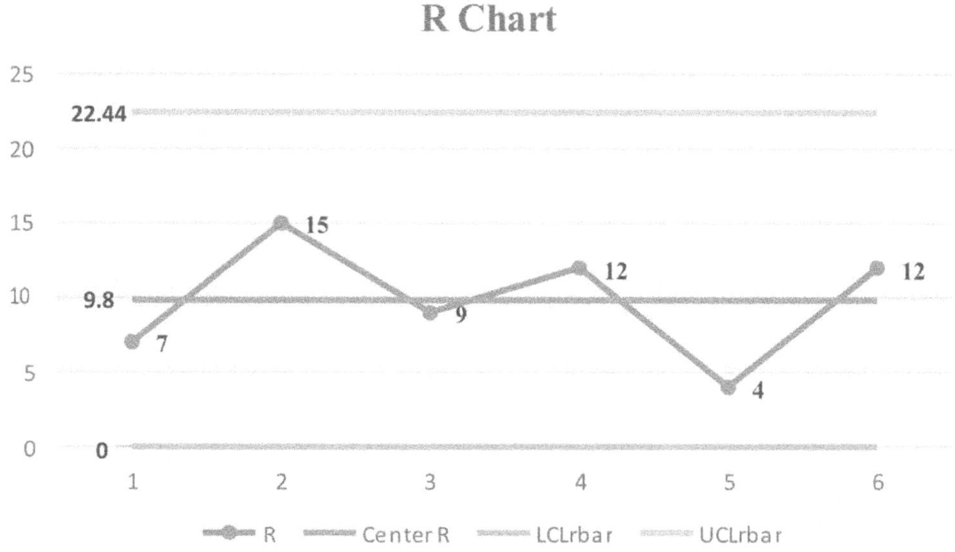

**For a more accurate representation see the electronic version.*

Figure 16. The process revisited after the changes were made during the Do stage

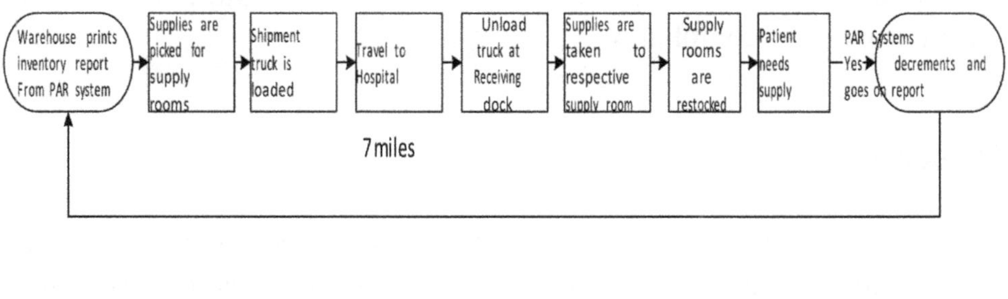

The largest cost was buying the bins for the supply room at a cost of $3395.04. After considering this for six different units the total will be $20,370.24. Management and the continuous improvement team justify this as a one-time cost due to the increased compliance and thus reduced lost chargeable items, the standardization across all supply rooms which in turns reduces the time to find supplies regardless of which floor a nurse works on.

The project is deemed successful and rollout begins across all nursing unit supply rooms.

A Gantt chart is created to line out the next few weeks of the project. The problems have been identified and defined as a whole and so each floor

Figure 17. As part of the Act phase a Gantt timeline for the next three units to be covered in this project is created

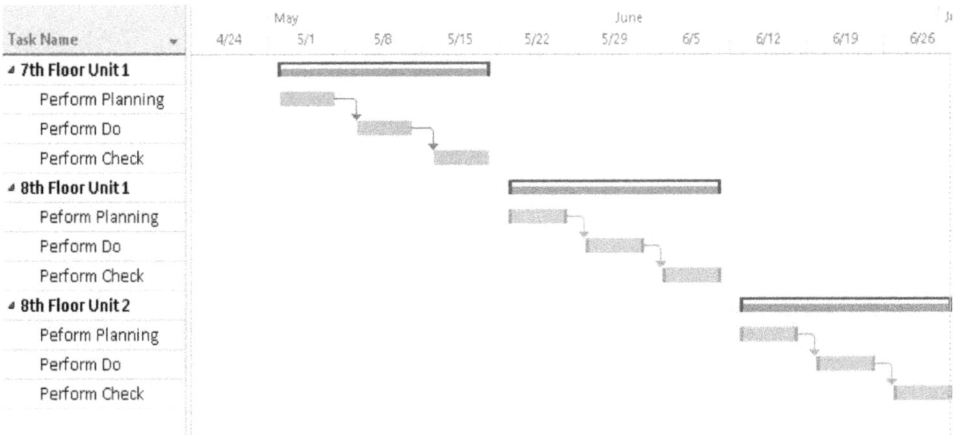

will begin with the plan step and end on the check step. Figure 17 shows the timeline for the rollout.

Although this unit, 7th Floor, Unit 2, has went through the initial IDPDCA, the continuous improvement cycle will continue to occur by performing a new Pareto chart and determining which item(s) are now causing the biggest issues for PAR compliance and the cycle will continue to address those items to further increase compliance.

CONCLUSION

This project was successful and completed within 10 weeks. Since the team now has experience and understands the process, this same process can be completed more quickly across other nursing stations. This is important because quality is now at the forefront within the healthcare industry. So much so that that patient satisfaction and other quality score are now how hospitals are reimbursed by Medicare (reimbursement). The industry is going to have to find ways to reduce costs while keeping quality high. This project successfully demonstrated that the quality philosophies and tools such as Lean and Six Sigma SPC can be adapted from the manufacturing sector to healthcare and other transactional process settings.

REFERENCES

Aikens, C. H. (2011). *Quality Inspired Management: The Key to Sustainability*. Boston: Prentice Hall.

Besterfield, D. H. (2013). *Quality Improvement* (9th ed.). Upper Saddle River, NJ: Prentice Hall. Print

Meyers, F. E., & Stewart, J. R. (2002). *Motion and Time Study for Lean Manufacturing*. Upper Saddle River, NJ: Prentice Hall.

Reimbursement, Linking quality to payment. (n.d.). Retrieved April 25, 2016, from https://www.medicare.gov/hospitalcompare/linking-quality-to-payment.html

Xbar. (n.d.). *What is an Xbar-R chart?* Retrieved April 10, 2016, from http://support.minitab.com/en-us/minitab/17/topic-library/quality-tools/control-charts/understanding-variables-control-charts/what-is-an-xbar-r-chart/

Chapter 5
Lean and Total Quality Management in Civil Engineering

ABSTRACT

The aim of this chapter is to examine the utilization of total quality management (TQM), ISO 9000, and Lean concepts in the management processes of civil engineering consulting firms. The history of management theory and the existing literature related to total quality management, ISO 9000, and Lean concepts as applied to the civil engineering profession were reviewed. Also, the practical application of these concepts in civil engineering firms' operations and the extent that academic courses in these topics are offered were analyzed. The study is based on internet research and interviews with civil engineering managers and college professors. The results show that while the literature strongly endorses the utilization of these management concepts, the actual use is little known in the professional business of civil engineering. Also, in general, undergraduate civil engineering academic programs in the United States do not offer specific courses, which cover these concepts in their curriculums.

INTRODUCTION

The issue is while TQM, ISO 9000 and Lean concepts have been suggested as a means to the improvement of civil engineering consulting firm operations; it appears these concepts are not being adopted by industry or academia. The purpose of this research is to determine if and possibly why the civil

DOI: 10.4018/978-1-5225-4062-5.ch005

engineering consulting profession is not, by practical application, adopting some form of TQM, ISO 9000 or Lean concept in management processes and what suggestions can be made to facilitate a change in the profession to consider or adopt these concepts.

Total Quality Management (TQM) has proven itself in the past to facilitate overall improvement of manufacturing firm process operations in many aspects and has more recently been recommended for application to engineering, design and construction firms. ISO 9000 and Lean philosophies are also suggested as ways civil engineering consulting firms can improve their organizations. Civil engineering has been especially considered relevant to Lean construction as applied to design and construction firms. This correlation of civil engineering to Lean construction is due to construction and civil engineering involving generally very similar projects, with civil engineering be more closely aligned with design, while construction with actual building processes. TQM, ISO 9000, and Lean concepts, which have been suggested as applicable to the civil engineering profession, have somewhat evolved, as have management philosophies over time.

Deming, Crosby, Juran and other management theory pioneers envisioned TQM initially as a means to improve manufacturing processes in Japan in the 1950's as the automobile industry developed internationally. The Motorola Corporation followed the initiation of TQM by introducing the concept Six Sigma, a philosophy aimed toward reducing defects in high volume manufacturing to a minimum in the mid-eighties. ISO 9000 followed in the late eighties and was created to establish a set of international standards by which firms could benchmark their abilities in comparison to other companies around the world (Davis and Goestch, 2006).

Lean theory, initially described as Just-in –Time manufacturing, essentially arose from the Toyota Corporation in the 90's through the work of Eiji Toyoda and Taiichi Ohno where they identified wastes in the manufacturing processes at Toyota Motor Corporation. They developed a system of production that reduced waste, however Lean theory was also later recognized as a fundamental business philosophy (Forbes and Ahmed, 2010). Lean theory was first recognized by Lauri Koskela as a new philosophy that could be applied to the construction industry and was later developed further by Glen Ballard and Greg Howell who coined the term "Lean Construction", they also initiated the Lean Construction Institute and pioneered the Last Planner System, all in management theory, to maximize value and minimize waste in the production of products and services in business organizations (Forbes and Ahmed, 2010).

ISO 9000 and Lean Management

Literature concerning TQM, ISO 9000 and Lean as management concepts has transcended over time with the advent of each of these philosophies development over time. TQM and ISO 9000 are basic to understanding the origins of the theory and implementation of management processes and well documented by Goetsch and Davis and (2006) in their reference text "Total Quality Management" which has been utilized since 1994 as a major study and reference guide. Generally, TQM and ISO 9000 are characterized by Goetsch and Davis (2006) as methods of improving any type of organization, suggesting that actual implementation of these concepts needs to be tailored to fit the specific organization's needs.

The focus of this study is the concepts of TQM, ISO 9000 and Lean applied to the civil engineering profession and although these concepts are similar in goals and objectives their differences are important. Goetsch and Davis (2006) differentiate TQM and ISO 9000 by describing TQM as an all-encompassing organizational process that not only includes the systems of design, methods of production and deployment of products and services, but also includes support systems such as human resources, finance and marketing. A TQM organization also requires total involvement of management and employees in a combined collaborative effort to assure the attainment of desired goals and objectives.

ISO 9000 is defined as being concerned only with the quality management systems for the design, development, purchasing, production, installation and servicing of products and services. The two concepts are not interchangeable and not in competition with each other. ISO 9000 can be viewed as a subset of TQM and ISO 9000 is frequently adopted in a non-TQM environment. Organizations can adopt either or both as they are compatible. ISO 9000 can be a head start on TQM (Goetsch and Davis, 2006).

In contrast, Lean is a production process considered a fundamental business philosophy, which views the expenditure of resources for any purpose other than the creation of value for a customer to be wasteful, it is lean because it uses less of everything (Anvari and Moghimi 2011), (Forbes and Ahmed 2010). The differences are that Lean focuses on improving entire value streams whereas TQM and like philosophies tend to focus on individual processes. Also,

TQM focuses on value-adding processes while Lean emphasizes eliminating waste and therefore adding value. Therefore, there is not a common point of similarity between Lean and TQM clearly acknowledged (Anvari and Moghimi 2011). Currently, Lean as applied to construction is in its early evolution and is slowly, but gradually being adopted (Forbes and Ahmed, 2011).

Several articles have been authored, specifically concerning the management of civil engineering firms, suggesting the utilization of TQM and similar concepts as a means of improvement. In general, management concepts have been surfacing as applied to engineering design and construction for some time, especially since Koskela introduced the connection of lean manufacturing to construction in 1992.

The recognition of a lack of understanding of the possible positive effects of modern management philosophies like TQM, ISO 9000 and Lean, as applied to civil engineering, is addressed in an article by Milorad Zlantanovic titled 'Management in Civil Engineering and It's Developing Trends" authored in 1999 in which he states:

Management in civil engineering is not applied seriously, which is an absurd considering significance and specific character of building industry for community, especially taking into account fight for survival and necessity of building company development.

Zlantanovic also addresses the lack of and importance of management education for civil engineering professionals, both academic and continuing education, which would assist to assure the promotion of modern management in the civil engineering profession.

In summary, the literature reviewed, both current and historical, emphasizes the management philosophies reviewed in this study may improve civil engineering design process management, however is not currently generally recognized by the civil engineering profession.

METHODOLOGY

This study involved researching information concerning the operations of civil engineering consulting firms and their overall management procedures to obtain an understanding of the basic nature of their business culture. Ten firms were selected at random from a geographic area covering the central United States representing a wide range of companies by size. The firms selected all provide civil engineering consulting services, with most firms providing general design of highways, site development, sewer and water and land surveying with the larger firms being more diversified and providing more of a selection of specialized services. Firms were initially researched on the internet to obtain basic available information with each firm later contacted individually by telephone. The individual at each firm, who was interviewed, were typically engineering managers or administrative staff knowledgeable of the firms' organization and management principles.

Specific information collected were years in business, geography of business services, number of employees, number of offices, types of services provided, management training programs provided for employees, award program participation, types of company manuals and knowledge and level of participation in TQM, ISO 9000 and Lean management philosophies. The results of the information collected are provided in Table 1.

Also, a cursory search of the availability of academic courses at several universities offering civil engineering undergraduate programs of study was conducted and several professors were interviewed to determine generally the basic level of management courses offered, especially those covering the topics of TQM, ISO 9000 and Lean in civil engineering programs.

DATA AND FINDINGS

The individual firm's researched ranged in size by employees from six to 12,000 with the largest firm being the only international firm interviewed. The other firms researched included one national firm, five multi-state mid-continent firms and three exclusive to Kentucky. The firms range in number of offices from 1 -200 with an average of 18 employees for each office location. The firms selected for study represent a common geographic area related to main office locations, however are greatly divergent by number of employees per firm in that one firm containing 12,000 is more than ten times larger than any of the other firms researched. One firm has 960 employees; five firms have between 80 and 200 employees, while the remainder has six to 12 employees. All of the firms have employees that work in the field. Listed below are summary statistics about the information collected from the sample of firms:

- 80% provide construction inspection services.
- 60% provide construction management services.

Table 1. Civil engineering firm survey data

Criteria	1	2	3	4	5	6	7	8	9	10
Years in Business	40	59	40	48	12	30	28		45	
Type of Services	Civil/Multi	Civil/Multi	Civil/Geo/Env	Civil/Multi	Civil/Multi	Civil/Multi	Civil/ Small	Civil/Small	Civil/Multi	Civil/Multi
Number of Employees	80	12,000	960	200	12	100	7	6	90	364
Number of Offices	(5) KY	200/INT	22/MID-US	10/MID-US	2KY	6	2KY	1KY	8/MID-US	21/NAT
Field Employees	Y	Y	Y	Y	Y	Y	Y	Y	Y	Y
Construction Inspection	Y	Y	Y	Y	Y	Y	N	N	Y	Y
Design Build	N	Y	N	N	N	N	N	N	N	Y
Testing of Materials	N	Y	Y	N	N	Y	N	N	Y	Y
Web Page	Y	Y	Y	Y	Y	Y	Y	N	Y	Y
Policy Manual	Y	Y	Y	Y	Y	Y	Y	N	Y	Y
Safety Training	N	Y	Y	Y	Y	N	N	N	Y	Y
Standards Manual	Y	Y	Y	Y	Y	N	Y	N	Y	Y
Statement of Qualifications	Y	Y	Y	Y	Y	Y	Y	N	Y	Y
ISO 9000	N	Y/ISO 9000 Cert	Y/F	Y/F	Y/F	N	N	N	N	Y/F
Lean Construction	N	Y/N	Y/F	Y/F	N	N	N	N	N	Y/F
TQM	N	Y/N	Y/F	Y/F	N	N	N	N	N	Y/F
Management Training	Y	Y/INTERNAL	Y/INTERNAL	Y/INTERNAL	N	Y	N	N	Y	Y
Awards	ACEC	ACEC	ACEC	ACEC	N	N	N	N	ACEC	ACEC

- One of ten provides design-build services and it is the largest firm which provides international services.
- 50% provide testing of materials.
- 90% have a company web page.
- 60% provide safety training for employees.
- 90% have a standards manual.
- 90% have a statement of qualifications.
- 70% provide some type of management training for supervisor employees, with 40% of the companies providing the training in-house.
- One firm is ISO 9000 certified, which is the largest firm interviewed.
- Three firms are familiar with ISO 9000, which are the largest firms of all others.
- 50% of the firms are not familiar with ISO 9000.
- 40% of the firms are familiar with Lean Construction, with none adopting the concept.
- 40% of the firms are familiar with TQM with none adopting the concept.
- 50% of the firms participate in an awards program, typical administered by an engineering society such as the American Council of Engineering Companies.

The other aspect of this research, related to the teaching of management concepts in civil engineering academia, included a cursory review of several civil engineering undergraduate programs geographically situated in the region of the where the subject civil engineering consulting firms are located. Information was obtained from the internet at university web sites and from interviews with several college civil engineering professors.

The general review of these program curriculums indicate general management courses for civil engineering are not required or offered in the undergraduate course programs. Interviews with several civil engineering professors reveal management courses related to civil engineering are most often taught in special study courses for seniors and graduate level classes, however generally none are strictly related to TQM, ISO 9000 and Lean construction. The professors indicate that the reason the courses are not offered is due to a lack of space for the courses in the programs of study and a lack of interest to develop management courses in undergraduate civil engineering programs, considering all other course requirements. Most often management engineering courses are offered and required in industrial and systems engineering programs of study, which are mainly manufacturing directed programs, or as special interest research topics of graduate engineering students in the various engineering fields of study.

ANALYSIS/DISCUSSION

Analysis of the information collected from the civil engineering consulting firms concerning business and management practices and relationship with TQM, ISO 9000 and Lean concepts reveal several generalities. However, the great diversity of the composition of the company's selected, especially number of employees, must be a consideration in the analysis of the data collected. The 12,000 employee firm reviewed is the only firm that provides international consulting services and office locations, design-build services and only firm that has adopted one of the management concepts focused on in this study, ISO 9000. The largest firm also generally adopts all the other business practices selected for study. The second, third and fourth largest firms, 200-960 employees, generally follows by adopting all the general business practices as the largest firm, except design-build services, and are familiar with the concepts of TQM, ISO 9000 and Lean construction, however have not adopted any of these concepts. The three firms ranging in size from 80 -100 employees follow most all of the general business practices however indicated in the interviews they are not familiar with TQM, ISO 9000 or Lean Construction. The smallest three firms, having 6-12 employees, follow some basic business procedures such as having policy manuals and standards manuals, however are not aware of the management concepts which are the subject of this study.

In summary of the analysis of the data collected, most of the firms with over 50 employees have some type of management training requirements and objectives; all of the largest four firms reviewed 364-12,000 employees, are familiar with TQM, ISO 9000 and Lean concepts and the only firm, having 12,000 employees, also providing international services, has adopted ISO 9000. ISO 9000 is generally considered a stepping stone toward TQM, however is considered exclusive of TQM or Lean management processes (Davis and Goestch, 2006).

CONCLUSION

In conclusion, based on the data collected, the civil engineering consulting firms that recognize and are familiar with the management processes TQM, ISO 9000 and Lean are the four largest firms, with the largest and only international firm having adopted one of the processes, ISO 9000. While this is a broad conclusion based on the small sample selected for review in this study, it appears analogous to an economy of scale being the relevancy of firm size related to degree of complexity of management style for the organizations studied. This is inherent of the management theories studied here, TQM, ISO 9000 and Lean concepts, for their historic origins are based on the problem of large company's striving to be successful by having to find ingenious ways to manage large numbers of employees in a very competitive international market place.

The literature reviewed in this study recognizes this issue of firms similar to civil engineering consulting firms failing to see the value of modern management styles such as TQM, ISO 9000 and Lean. Proponents of TQM, ISO 9000 and Lean support the movement of organizations toward recognizing, learning about and adopting these concepts as they may be applied to their specific organization and needs. This is accomplished by suggesting and providing the opportunity for education and training of these management concepts at both the professional and academic level.

Unfortunately, until education of the management concepts researched in this study become part of undergraduate civil engineering programs of study, the recognition of the importance and application of theories like TQM, ISO 9000 and Lean in civil engineering will be slow. All states in the United States now require a college degree to become a licensed professional engineer, therefor the opportunity exists for consulting civil engineers to become aware of the benefits of management theories for the improvement of the profession if courses are offered covering these concepts. Also, most states now require licensed professional engineers to have continuing education training and thus providing an opportunity for spreading the word of modern management styles such as TQM, ISO 9000 and Lean Construction. History tells us that modern management is known to improve an organization no matter size or type of organization as long as it is applied to fit the needs of the individual firm.

Future areas of study include researching a larger sample of firms having over 500 employees in regards to their recognition and adoption of TQM, ISO 9000 and Lean management concepts and to delve into the application of Lean Construction principles in civil engineering due to their similarities of types of projects and interconnectedness.

REFERENCES

Ambrose, G. A. L., & Claasen, S. J. (2004). *A Total Quality Management Approach to Civil Engineering Consulting*. Retrieved from http://hdlhandle.net/2263/20518

Fedderle, M., & Chase, G. (1993). Appling Total Quality Management to Design and Construction. *Journal of Management Engineering*, *9*, 357–364.

Forbes, L. H., & Ahmed, S. M. (2010). *Modern Construction: Lean Project Delivery and Integrated Practices*. Taylor and Frances.

Goetsch, D. L., & Davis, S. B. (2006). *Quality Management*. Pearson Prentice Hall.

Lam, S., & Tang, C. (2002). Role of Surveyors under ISO 9000 in the Construction Industry. *Journal of Surveying Engineering*, *128*(4), 187–199.

Lam, S., & Tang, C. (2004). A TQM Model for Construction Surveying Under ISO 9000. *Geomatica*, *58*(3), 195–202.

Lam, S. Y. W. (2010). *TQM of Engineering Survey Operations under ISO 9001 in Hong Kong*. FIG Congress 2010. Facing the Challenges- Building the Capacity, Sydney, Australia.

Morehead State University Webpage. (n.d.). Retrieved from http;/www.moreheadstate.edu/content

Murray State University Webpage. (n.d.). Retrieved from http:/www.murraystate.edu/academics

Oakland, J.S., & Aldridge, A.J. (1995). Quality Management in Civil and Structural Engineering Consulting. *International Journal of Quality & Reliability Management, 12*(3), 32-48.

Purdue University Webpage. (n.d.). Retrieved from http: /www.purdue.edu/academics

Virginia Tech Webpage. (n.d.). Retrieved from http://www.vt.edu/academics/course_listing.html

Zlantanovic, M. (1998). Management in Civil Engineering and Developing Trends. Facta Univeitatis. *Series Architecture and Civil Engineering.*, *1*(5), 637–644.

Chapter 6
Single-Minute Exchange of Dies at a Kaizen Event

ABSTRACT

Lean concepts play a fundamental role in the promotion of continuous improvement in the workplace. This chapter seeks to assess the impact of Kaizen events on an organization's bottom line, in the case of Single-Minute Exchange of Dies (SMED) that is a system for dramatically reducing the time it takes to complete equipment changeovers. In this case study, SMED was used in combination with other lean tools such as 5S and standardized work in a bid to promote continuous improvement of the "flawed" operations that occur in a world-class manufacturer plant. This analysis is critical in determining whether the path to achieving continuous improvement process for the team in the organization can be achieved through the use of SMED. The results have shown that a lean strategy like SMED, coupled with other lean strategies like 5s and TPM, plays a fundamental role in reducing process inefficiencies in the plant.

INTRODUCTION

Single-Minute Exchange of Dies (SMED) is a critical lean production method that is concerned with the reduction of waste in a manufacturing process (Dave & Sohani, 2012). The application of SMED is aimed at the achievement of the optimization of machine utilization, reducing the production times, stocks, and machine adjusting times, as well as the time that the machine

DOI: 10.4018/978-1-5225-4062-5.ch006

is not operational, and it also works towards having small lot sizes (Ulutas, 2011). The concepts that are critical to SMED are put to work when trying to achieve a reduction of the setup times. An analysis of the set uptime is the first step that is done to assess the current state. This is followed by the elimination of the inefficiencies that are present as far as time is concerned. The next step then requires the identification of the internal and external setups, after which, the internal setups are converted to external ones. The final step requires standardization, where improvements were identified and plan to eliminate process wastes are provided. In combination with other lean tools like 5s, TPM, and standardized work, more as far as process improvements can be achieved with the application of SMED.

The essence of the SMED system is to convert as many change-over steps as possible to "external" (performed while the equipment is running) and to simplify and streamline the remaining steps. SMED can be said to be a process-based innovation that was brought to the publishing scene in the mid-1980s (Carrizo & Campos, 2011). SMED plays a fundamental role in instigating productivity in the firm (Carrizo & Campos, 2011). SMED is essential when it comes to converting a manufacturing process from running one product to running the next, in what is termed as a rapid change-over (Dave & Sohani, 2012). The rapid change-over is imperative in reducing the lot size and improving the production flow (Dave & Sohani, 2012).

When narrowing down to looking at what SMED in fact means, it is essential to note that the term is drawn from the goal that is concerned with the reduction of the changeover time, to the single digits, less than ten minutes to be precise (Vorne, 2016). This is possible, especially where the elements of a given changeover are analyzed to ensure that they can be eliminated, streamlined, moved or even simplified (Vorne, 2016).

One aspect that stands out in SMED is the essence of a quick changeover, which mainly captures the concept of time. It is essential to address what is, in fact, meant by the term "quick changeover." A quick changeover is a process that is aimed at reducing the time that has elapsed from the production of the last good part to the production of the next good part after a set-up (Manufacturing Terms, n.d). In actuality, a quick changeover is an essential concept given that it allows the necessary flexibility that matches the production mix, to the existing demand (Velaction Continuous Improvement, 2016). Quick changeovers facilitate the production of smaller batches, and the presence of continuous flow (A Lean Journey, 2010).

The role of SMED in lean manufacturing is undisputed, particularly when looking at the concept of Total Productive Maintenance (TPM). As such, it is imperative to look at the essence of TPM in lean manufacturing keenly. TPM is fundamental to the implementation of lean manufacturing (Lean Manufacturing Tools, 2017). Note that process improvement is facilitated by the use of machines and equipment; therefore, the role of TPM can only be promoted with the reliance on machines (Lean Manufacturing Tools, 2017). When combined with 5S, TPM provides the basis for the improvement of business operations (Lean Manufacturing Tools, 2017).

TPM makes use of SMED in the pursuit of increasing availability of the Overall Equipment Effectiveness OEE (Weigh Label, 2017). The alignment of SMED and TPM priorities allows the understanding of OEE losses. TPM is necessary for the company's bottom line given that it fosters critical improvement of the maintenance functions that take place therein, and which requires the involvement of human resources (Al-Hassan, Chan, & Metcalfe, 2000). As evidenced by Al-Hassan, Chan, and Metcalfe (2000), when it is implemented correctly in an organization, TPM precipitates an improvement in the quality, productivity, and a reduction of unnecessary costs. As a matter of fact, with the increment in the degree of automation in an organization, then chances tend to be high that a greater cost reduction that is generated from TPM occurs.

Financial implications that are attributed to TPM cannot be disputed. In reality, organizations should be aware that before TPM implementation, costs and benefits amounting to the lean concept ought to be understood (Smartware Group, 2016). It should be noted that costs that are attributed to time and human resources cannot be avoided (Smartware Group, 2016). When organizations stick through with the expenses as mentioned above, they can, in turn, prove to be productive (Smartware Group, 2016). Some of the costs take note of training of the manpower, as well as the expenses incurred by additional maintenance (Smartware Group, 2016). These costs can, in fact, result in budget overruns, but it is essential that organizations maintain their focus, with the aim being centered on achieving improvements on the OEE, within every given year of TPM implementation (Smartware Group, 2016). Even with the costs that are likened to TPM implementation, financial benefits can be achieved in the long run. Organizations are likely to reap benefits in areas such as long equipment life cycle, mitigated logistic hurdles for employees, and the presence of maintenance challenges that are much simpler (Smartware Group, 2016).

In summary, it is essential to consider that SMED and TPM are critical concepts that drive other lean tools towards the achievement of success for an organization. For instance, 5S and standardized work can both be achieved with SMED and TPM. 5S ensures that tools are located in their place before a changeover can take place, or when a maintenance duty is ongoing (Wrye, 2011). The result is that during both processes, the machine is down for a short time before production can be given a go-ahead to commence (Wrye, 2011). Standardized work, on the other hand, promotes the efficiency of changeovers, making them faster and safe (Wrye, 2011).

The initial state of work at the plant that was studied revealed the fact that the loopholes that were in the organization. In fact, the current state of quality management in the organization was not up to par with organizations that have working systems. The plant had Standard Operating Procedures, meaning that it was aware of this part of the management systems, but its contents were not well defined for the understanding of the user. Additionally, the initial state indicated that the firm lacked a 5s protocol, which is evidenced by the poor organization skills that were evident in some areas of the plant. The plant also failed in its sustainment procedures, a scenario that is expected given that it had deficiencies in the two critical components (SOPs and 5s) that were outlined earlier. From this information, it is apparent that the plant is in dire need of having its quality management systems in order, by the implementation of lean strategies (SMED in conjunction with 5s). This will go a long way for the firm, as far as eliminating wastes is concerned.

LITERATURE REVIEW

Standards and Davis (2000) breaks down lean into three parts: lean philosophy, lean principles, and lean practices. Lean philosophy focuses on "total system efficiency, continual improvement, value-added activity, and respect for people" (Standards & Davis, 2000). Lean principles emphasize the reorganization of" the flow of production material throughout the entire enterprise" (Standards & Davis, 2000). This is supported by lean practices through decreasing the variability of production.

In order to make lean applicable, an organization must not include one of the previously mentioned elements, but all of them. The research Bhasin (2011) conducted "attempts to establish that Lean is a total system supporting and encouraging its employees to continuously improve the processes they work on" (Bhasin, 2011). This goes back to why communication and its factors are important. Continually carrying out an efficient and accurate system of communication will make the lean system beneficial. When viewing and implementing lean, Bhasin (2011) mentions that one should not look at it in a narrow sense but analyze the entire organization.

For a business to be able to embrace being lean, it must consider the above mentioned. The organization must adopt being lean into its company's practices, vision, and mission. Every aspect and division of the business must operate based on being Lean, including outsourcing.

In the study Bhasin (2011) conducted, he discovered that the organizations that displayed the best performances involved: "higher profitability; improved employee performance; improved market share; increased competitiveness; and constant waste reduction."

"The baseline is that markets, technology, people, and money all need to be attended to for a business to be successful" (Bhasin, 2011). The companies used in Bhasin's study used Lean as a business philosophy, thus demonstrating the above results. This leads to the implication that applying the lean philosophy to an organization can better the company in all areas that judge a company's performance level. Stone (2012) suggested that the definition of lean should consist of three terms that describe the end of the process, what occurred to achieve the ending goal, and what was used to carry out the process.

Bhasin concluded that lean would make planning and processing easier while shortening the lead time through the supply chain. He also found that Lean should be "viewed as a developing discipline and dynamic since it is continuously improving" (Bhasin, 2011). Organizations should use Lean as a "long-term commitment" (Bhasin, 2011) that is both physical and financial. The more it is practiced, the simpler it would be to lead the business into obtaining its benefits.

Lean Manufacturing

The idea of lean production did not begin with the Toyota Production System (TPS) or the Just in Time Production (JIT). It began as far back as Eli Whitney's interchangeable parts of the cotton gin in 1799 and the muskets the thousands of muskets he manufactured for the U.S. Army at an unbelievably low price. Fredrick W. Taylor developed the idea of Scientific Management, which led to Lillian Gilbreth studying workers and involving psychology. This began the concept of eliminating waste. Then came Henry Ford who became known as "the first practitioner of Just in Time and Lean Manufacturing" (Sun, 2011). It wasn't until the 1950s and 1960s that Eiji

Toyoda and Taiichi Ohno of Toyota began developing more approaches to lean manufacturing (Upadhye et. al., 2010).

In the late 1990s, the introduction of Lean Thinking encouraged organizations to come to a better understanding of planned change throughout the organization as a whole (Stone, 2012). According to Standard and Davis (2000), productivity has risen by 35% between 1995 and 2000 and is continuously increasing. In order for manufacturers to keep up with competition, they must make sure they are continually improving.

Steinlicht (2010) classifies the techniques of lean into three views. He calls the first view "benchmarking," or imitating other organizations to keep up with the competition. The second view is called "bolt-on," where an organization adapts its local conditions to lean principles. The third view is called "adaptation." This means to "completely reinvent the organization (using the principles of lean instead of the tools and practices)…This approach uses lean to achieves cultural change" (Steinlicht, 2010).

Stone (2012) suggested that the principles of lean are related to the "tools used to execute" the process. Kovács (2012) identifies five principles of lean manufacturing: "Value, the value stream, flow, pull, and perfection.

"Some lean organizations also rely heavily on Six Sigma, which includes a set of statistical techniques for identifying and reducing process variation" (Sun, 2011). Krause (2009) states that Lean Six Sigma focuses on: "Delighting your customers with speed and quality; improving your work processes; working together for maximum gain; basing decisions on facts and data."

Compared to the basics of Lean philosophy itself, lean production and manufacturing not only emphasizes increasing profits, competitiveness in the global market, and employee involvement, it focuses on its customers as well.

According to Kovács (2012), applying lean manufacturing comes with many advantages such as short production lead times, operation timing imbalances are more obvious, mistakes become more obvious, and their source can be identified quickly, there is a continuous drive to improve, and part handling and inventory costs are at a minimum. The transformation to using a lean strategy has been proven to be more successful when the entire organization is aligned with the same strategy (Stone, 2012).

Lean manufacturing has been shown to improve the competitiveness of manufacturing organizations. The concept of lean manufacturing started with the Toyota Production System. Then, the idea was expanded by a research group at Massachusetts Institute of Technology, starting in the 1980s (Womack, Jones, & Roos, 1990). The philosophy of lean manufacturing is to use a process of waste reduction, thus producing higher quantities and better quality products with the same or fewer resources. The goals of lean manufacturing are zero lead time, zero inventories, and zero defects, resulting in higher customer satisfaction (Tapping, Luyster, & Shuker, 2002).

Lean manufacturing means continuous improvement towards an ideal through the relentless reduction of waste, where ideal means delivering what the customer requests on time, on demand, and free of defects (Miller, 2005). The process can be traced back to the early 1900s, when in 1913; Henry Ford introduced the concept of mass production (Miller, 2005). Taiichi Ohno at Toyota was the first to present lean manufacturing in its current term to eliminate production waste (Soderquist & Motwani, 1999). Types of waste include mistakes which require recertification, production of items no one wants, process steps that are not needed, goods or employee movement without any purpose, goods, and services that do not meet the needs of the customers, and any waiting due to blockages or shortages of product. Taiichi Ohno classified these wastes into seven basic types: overproduction, transportation, process waste, operator movement, and inventory, idle time, bad quality (Bateman & David, 2002). Continuous improvement through attacking the seven wastes is a critical aspect of lean manufacturing. Table1 shows different issues that will be seen when the current state VSM is completed. The table also shows what lean tools can be used to help overcome these issues.

The Importance of a Team for Kaizen

Lean manufacturing is team focused. It requires knowledge to be pulled from everyone (from the hourly worker to the upper management), and it is driven by continuous improvement (Kaizen). The team focuses on continuous improvements in lean manufacturing and uses value-stream mapping as a tool to identify wastes. Harris and Donatelli (2005) state that "value-stream mapping is the foundational tool used by any company that is on the cutting edge of transformation from a traditional organization to a lean enterprise." Value-stream mapping of products was started through the Toyota production system of lean manufacturing. Ninety-four percent of manufacturing errors or problems belong to systems as per Deming (1986), lean manufacturing attacks these systems with the common goal of cost reduction or improvement of production (Deming, 1986).

Training is an important concept in Lean. Called Kaizen, it's a Japanese word meaning continuous improvement. Kaizen is a Japanese word made up of two words, "Kai" and "Zen." Kai means thinking, and Zen means good (Mika, 2005). Together they mean continuous improvement. Kaizen event is a planned, organized event that is made to challenge everyone in an organization working together to make improvements without massive capital investments. Team members immediately apply what they learn about Lean and Six Sigma to solve problems. Kaizen events are a way of accelerating improvements in worker productivity, helping management find new ways to gain substantial savings in time-space and labor output. Worker ideas are highly encouraged. Frequent or small improvements will result in shrunken lead times, a dramatic reduction in work-in-process, and reduces scrap and defects while minimizing the need for capital expenditures (Mika, 2005).

Kaizen events are important because they provide an excellent return on investments of financial and human resources. And, since Kaizen never actually ends, continued improvements will compound the return (Mika, 2005). Kaizen events often eliminate the need for costly overtime by improving process while collapsing lead times and dramatically reducing work-in-process. The importance of this event is to help focuses on improving material flow, information flow, and process quality of business.

Training in lean is of particular significance today, and it is considered to be a winning strategy. Sometimes companies cannot manage the competition and are forced to lower the charges and lose profit. While doing this, they might also lose clients, so sometimes the strategies applied only lead to disaster. The most profitable thing companies should do when they have problems is implementing a lean strategy. A lean strategy has proved to be one of the greatest advantages when it comes to business. Some of the primary reasons why a company should apply lean are to compete effectively in today's global economy and continued focus on the marketplace on quality, cost, and on-time delivery (Alukal & Manos, 2006). To compete successfully in today's economy business need to be at least as good as any of their global competitors, if not better. This goes not only for quality, but also for costs and cycle times (lead time, processing time, delivery time, and setup time). Lean emphasizes teamwork, continuous training, and learning, production to demand (pull), mass customization and batch-size reduction, cellular flow, quick change over, and total productive maintenance (Alukal & Manos, 2006). Not surprisingly, lean implementation utilizes continuous improvements approaches that are both incremental and breakthrough.

METHODS

Settings

The study was conducted at one of the plants of a world-class manufacturer of signal transmission products primarily for the entertainment, residential, industrial and security markets. The company is one of the largest U.S. based manufacturers of high-speed electronic cables and focuses on products for the specialty electronics and data networking markets. The company reports in the following four segments: the Americas segment, Europe, Middle East, and Africa (EMEA) segment, the Asia Pacific segment, and the Wireless segment.

The Americas segment contributed approximately 54%, 52%, and 57% of the company's consolidated revenue in 2009, 2008, and 2007, respectively with an industrial market to include applications ranging from advanced industrial networking and robotics to traditional instrumentation and control systems. Their cable products are used in discrete manufacturing and process operations involving the connection of computers, programmable controllers, robots, operator interfaces, motor drives, sensors, printers, and other devices. They also cover industrial environments, such as petrochemical and other harsh-environment operations, which require the cables with exterior armor or jacketing that can endure physical abuse and exposure to chemicals, extreme temperatures, and outside elements.

Methodology

The methodology that was employed in the study took note of the use of time observation, where an assigned time was implemented on each given element. A spaghetti chart was also used to undertake an evaluation of the current process. Finally, photo and video documentation were done, where before and after photos were taken, with the aim being to capture the improvement of the entire process.

Team Members

The concept of the "team" plays an important role in lean manufacturing. A lean team consists of individuals who have the empowerment to instigate positive change in the workplace, thus promoting problem-solving promptly

(Bradbury, 2017). Teams tend to be formed by current processes, as pointed out by Bradbury (2017). In the present study, the team constituted individuals that held the following titles;

Engineer

The Collins English Dictionary (n.d) defines an engineer as an individual that makes use of scientific knowledge to design, maintain or construct machines or other structures. An engineer facilitates the repair of a mechanical or electrical gadget.

Maintenance

A maintenance personnel is concerned with offering support to maintenance operations at a given workstation.

Line Supervisor

This individual plays a critical role in the supervision of the workers undertaking operation and production duties.

Machine Operator

This individual has vital skill in the operation of a given kind of industrial machinery.

Tool Operator

A tool operator works with or tends to a particular set of tools.

Instructor and Facilitator

Promotes the action of teaching or transferring knowledge to others.

Figure 1 shows a group of the Kaizen Team that constituted of the engineer, maintenance personnel, line supervisor, machine operator, tool operator and the instructor/ facilitator.

Figure 1. Group Picture of the Kaizen Team

Problems Identified in the Original Process

Certain flaws were identified in the original process. Some of the flaws took note of the fact that the setup time was deemed to be excessive. The duration took about 153 minutes. The original process also revealed the fact that there were wastes in the setup steps. The steps totaled to over 1737 spaces. Another problem that was detected was the lack of 5S as well as the absence of clearly defined Standard Operating Procedures (SOPs).

Standard Operating Procedures (SOPs)

Standard Operating Procedures or SOP's are a set of clearly written instructions, which outline the steps or tasks needed to complete a job, operation or operate a piece of machinery or plant (Standard Operation Procedure, 2009). The Standard Operation Procedure (SOP) is a useful part of the quality system as shown in Picture 5. The purpose of an SOP is to carry out the operations correctly and always in the same manner (Standard Operation Procedure, 2011). This task enables anyone to understand how to perform the job, or the duties therein. The instructions must be written briefly in simple language that all operators and employees required to carry out the task can read and understand (Standard Operation Procedure, 2009).

In addition to the absence of SOPs, a sustainment process was also absent. It should be noted that in the bid to achieve total lean sustainment, it is essential that organizations leadership and employee engagement must be present (Gregory, 2012).

Goals

The goal of the study was to assess performance metrics through a set of Kaizen events at a specific machine station in the plant. During the week-long intervention that preceded a final report-out session, brain storming sessions, observation, and concept knowledge assessment was used to evaluate and increase productivity in the targeted station.

The tasks consisted of the machine operator to work with the team described above to fulfill the following outcomes:

1. To understand the elements that are involved in the setup the team evaluated each step in the process and developed a breakout of the categories
2. To use a video of a set up to detail each element therein
3. To assign time to each of the elements and use sticky notes to line up the setup elements visually

Tools Used

The following tools were critical to the present study;

Observation

Direct observation is a way through which evaluation information can be collected (Taylor-Powell & Steele, 1996). Observation facilitates the documentation of activities, without depending on people to respond to questions (Taylor-Powell & Steele, 1996). Observation is a clear way of obtaining direct information, such as the procedures that take place in a given firm or the management operations that are in place. Time observation falls under the bracket of direct observation.

Time observation allows for the development of standard work. This activity is essential for the determination of the cycle times of the steps in a given standard work (Velaction, 2016). This routine aspect of lean calls for multiple observations; in the bid to identify the most accurate cycle times (Velaction, 2016).

Brainstorming

Brainstorming is a technique that is used to generate ideas from group members, around a given area of interest (Infinite Innovations Ltd, 2011). The reason why brainstorming is preferred is that it allows easy generation of many ideas with little effort (Infinite Innovations Ltd, 2011).

Figure 2 shows the Kaizen team going through a brainstorming session, in an effort to generate ideas on how specific processes in the plant can be improved.

Spaghetti Diagram

A spaghetti diagram, as outlined by Bialek, Duffy, and Moran (2017) is a visual representation that traces the path of a given activity or an item, by the use of a continuous flow line. The spaghetti diagram plays a fundamental role in the identification of any redundancies in the work flow and the opportunities that can be employed to promote process flow (Bialek, Duffy & Moran, 2017).

Figure 2. Team Brainstorming Session

Critical intersections are identifiable, with the use of a spaghetti diagram, which signifies the areas that cause delays in a process. It is essential to note that waiting is a critical waste in lean, and so a spaghetti diagram comes in handy to reveal this waste and what can be done to mitigate or eliminate it.

Figure 3 shows a spaghetti diagram of the event. The red lines show the continuous flow lines, which are critical in the identification of the flaws of the process and how they can be addressed.

Additionally, in a Kaizen event, there is a need for having a definition and implementation of the measures of performance. Measures that show the past and the current levels of performance can be facilitated through the use of photographs and videos, which compare before and after performances after a Kaizen event has occurred (Lean Manufacturing Tools, 2017).

Figure 3. Event's Spaghetti Diagram

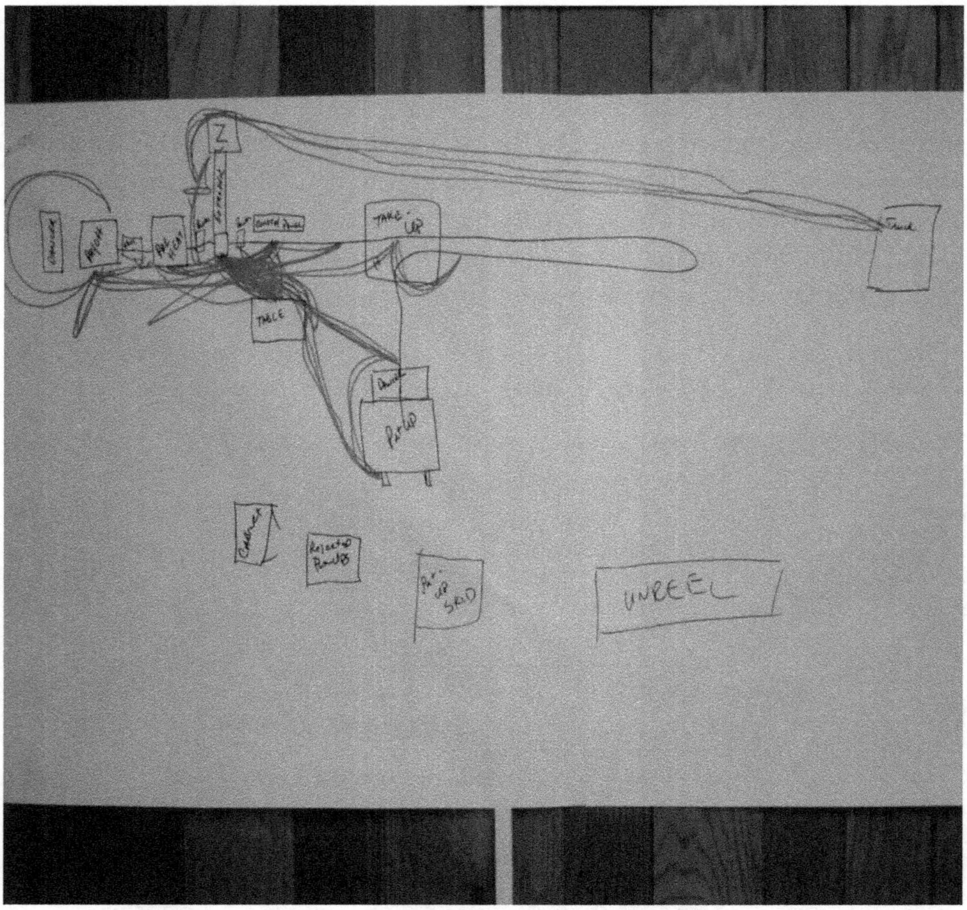

Current Process

Mostafa, Dumrak, and Soltan (2013) identified that lean manufacturing that is said to be successful is one that is introduced as a road map. On that note, lean assessment has to take place, where the current state of a process/activity is identified, and a map for the future state is also determined before implementation takes place. In the present project, the current process was identified by the use of observation, brainstorming, and a spaghetti diagram.

Under observation, time observation was considered, where identification of the cycle times was the ultimate goal. The current state of the cycle time would, in turn, facilitate the identification of the expected future state.

Through brainstorming, the member of the lean team can identify ways of promoting lean manufacturing in the plant, particularly when looking at lean concepts such as SMED, TPM, and 5S, and the determined current state. Finally, the spaghetti diagram will also be essential in the determination of the state current processes, and this will be important in determining the areas that allow the development of wastes and how they can be mitigated.

RESULTS AND ANALYSIS

Concept Knowledge

The participants were given three assessments to evaluate their knowledge of basic lean principles. The assessments were given in the form of a simple set of questions during the first day, the third day, and last day of the event. The results show that the participant performance as a group improved incrementally with time. The Table 1 shows the test score of the participants' performance during the three different times of the week. In the first trial, the score was at 23%. The results of the second trial were 68%, while the third trial was at 83%. By the time the participants were getting to the third

Table 1. Knowledge Performance Score

Trial	Score (%)
1st	23
2nd	68
3rd	83

trial, an improvement of 60 percent was noted. This confirms that training was essential for promoting lean manufacturing in the plant.

Improved Metrics

A measure of the results obtained from the study was done in the bid to assess the level of improvement. The results show that full change over time improves by a third compared to before the Kaizen event. Note that changeover time refers to the time it takes the production of the last good part to the production of the next good part after a set-up (Manufacturing Terms, n.d).

Table 2 shows a representation of the Kaizen metrics, where the performance of the operators was to be analyzed. In order to determine the changes in the performance level, there is the need for having the section that assesses the KPI measures before and after Kaizen event, with a clear identification of the expected target at the end of the event.

Test on the Setup

A test on the setup was done, to initiate changes in an effort to improve the current process in the plant.

Figure 4 shows the number of operator's tasks that are to be considered by the respective team member in order to propagate the test on the setup.

Table 2. Before and after Operator's Performance

Kaizen Metrics									
Metric (TTI or KPI)	Before Kaizen Baseline	End of Kaizen Target	End of Kaizen Actual	Improvement %	Sustained Target	1 Month Out Actual	2 Month Out Actual	3 Month Out Actual	Sustained % Improvement
Full changeover (per video)	1hr 50min	30 min							
Screw pull	??	20 min							
Color change	??	7 min							
Setup between 2 online jobs	??	30 min							
Compound scrap per set up	??	<= current scrap							
5S score for PE32	??	2x current score							
Prework Data/Information Collection Needed?									

Figure 4. Number of Operator's Tasks

The original setup was identified to be 153 minutes. The new setup was determined to be 28 minutes. When assessing the difference between the original and the current setup times, it was established that a 446% improvement level was attained.

Process Changes

An assessment of process changes was facilitated by the creation of a spaghetti chart. The chart was employed to ascertain the state of the current process. It was established from the chart that originally, 1737 paces were achieved. Figure 5 shows the current process in the plant, as evidenced by the spaghetti chart.

The original 1737 paces were then reduced to about 500 paces, with the employment of other lean tools, which were aimed at continuous improvement. The paces were then reduced from 500 to the current state of 255. This is as indicated in the spaghetti chart that is outlined in Figure 6.

Figure 5. State of the System before the Event

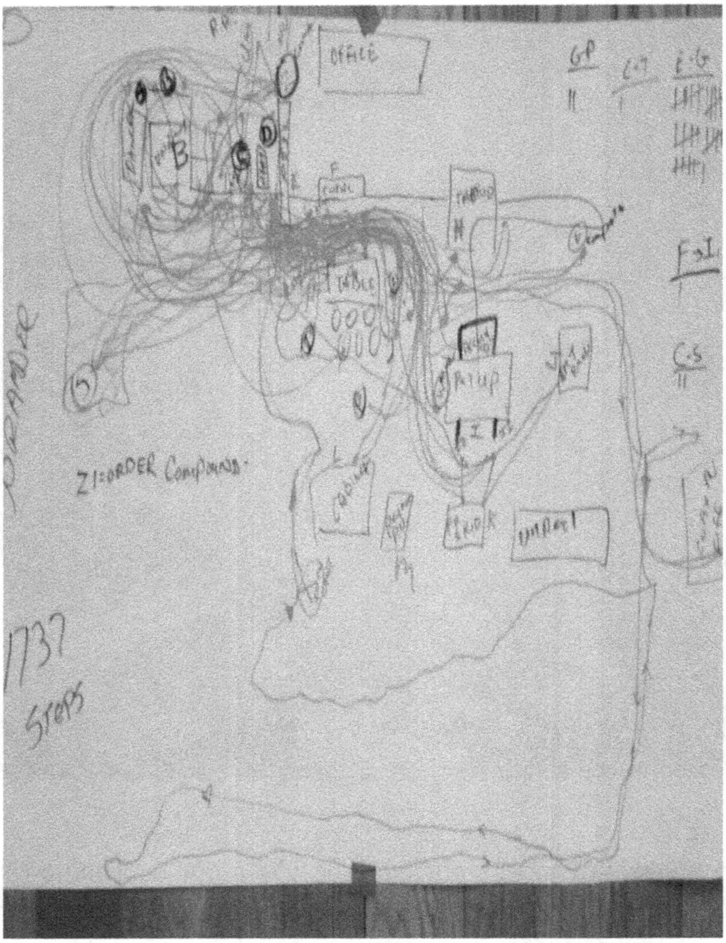

Sustainment: Lean Enterprise

Womack and Jones (1994) pointed out that when one member of a team makes progress in lean while another falls short, it is highly likely that such a time would not reap the full benefits that are attributed to lean. That said it is essential to note that in the bid to have sustainment in a lean enterprise, it is important that various kinds of needs are addressed, which can be in the form of the needs of individuals and those of functions (Womack & Jones, 1994). Sustenance allows instilling discipline such that the chances of backsliding are eliminated (Lista International Corp, 2017).

Figure 6. State of the System after the Event

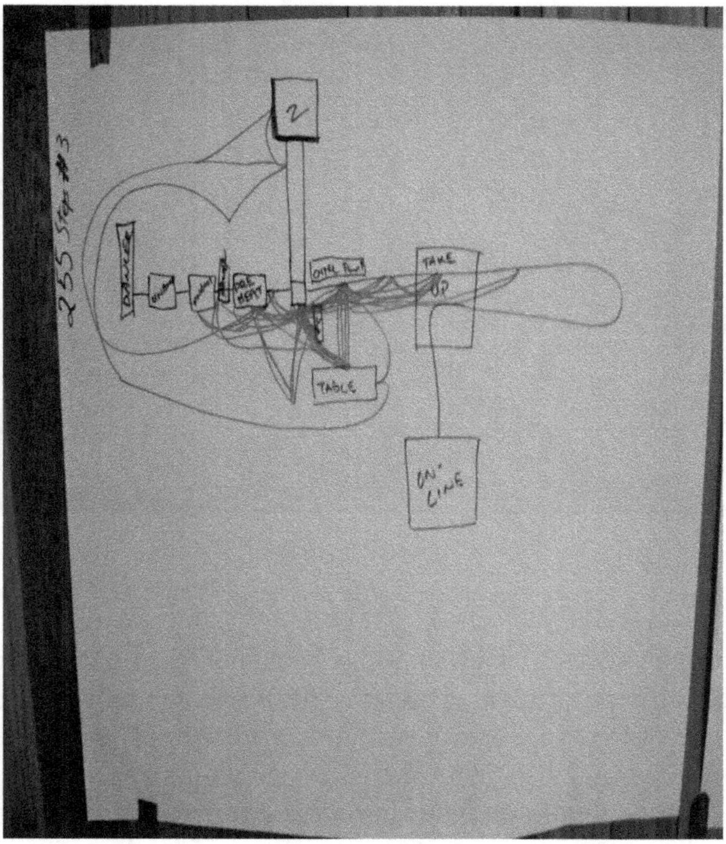

With the goals outlined earlier in place and armed with the tools identified above, the following were observed;

The overall set up time was noted to be at 28 minutes, while the overall steps as indicated in the spaghetti chart revealed the presence of 255 paces. A tool board was built and installed. 5S was also basically implemented, backed up with the presence of a sustainment process. A defined SOP was also in place, together with a sustainment process

Furthermore, training employees on the 5S activities were done. Here, the five pillars that make up 5S i.e. sort, set in order, shine, standardize and sustain were addressed. Figure 7 shows photographs of the before and after performances after the 5S activities were taught.

A review of the 5S sustainment was done, where the aim was to maintain a habit that fostered maintenance of correct procedures. Visuals of changeover

Figure 7. Showing the Before and After Images of Tool Organization

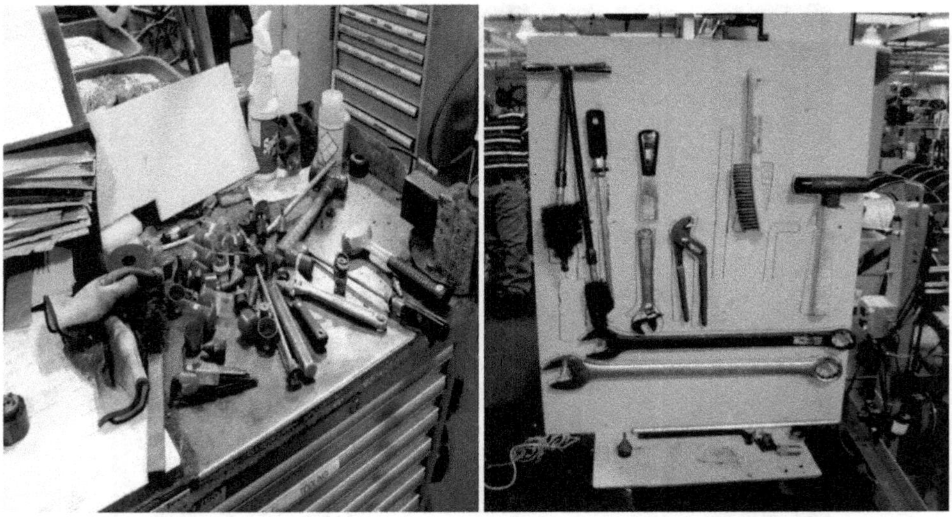

Before **After**

steps were also created, which played a fundamental role in improving the working conditions during the process. This was essential for the reduction of waste and down times that were unplanned. An analysis sheet for the setup operations was also finalized. This operations analysis sheet would be essential in facilitating monitoring activities that would make sure that conditions do not deteriorate to the original state. Finally, safety and waste issues were corrected. It is essential to note that 5S lean method is fundamental in encouraging the staff members to improve their work conditions. That said, safety and ergonomics are considered. With the improvement of the work conditions, then safety is likely to be guaranteed, as well as reduction of waste. Figure 8 shows a photograph that identifies the before and after of a work area.

The "before" image appears to be chaotic, while the after one showcases some form of organization, which promotes the essence of a safe and efficient work area, which in turn promotes SMED during a changeover.

CONCLUSION

The purpose of a lean system is to increase the throughput of a company to meet customers' product demands and to provide them with what they

Figure 8. Showing Additional Before and After Images of a Section of the Plant

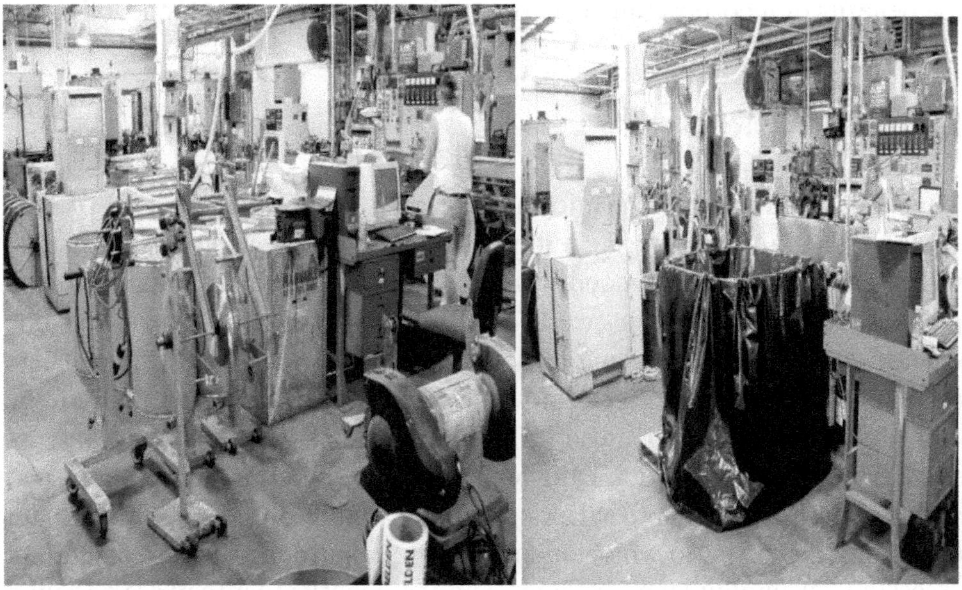

Before **After**

want when they want, and at the same time meets the customers' quality expectations. In a global supply chain environment, the identification of the major performance measures and their assessment is critical for firms to continuously improve their processes. The results of the study showed that the reliance on systematic strategies using lean tools and principles, and the emphasis on training are the key element to a successful lean deployment in a global supply chain environment. It is notable that a positive impact is felt in the work area, especially with the presence of teamwork. It is also evident from the study that the layout organization of the plant plays a crucial role in promulgating success in a given lean effort. It is highly likely that with the sustenance of the lean principles, SMED can be achieved in the plant under study. The changes that have been evidenced by the use of the tools identified earlier confirm that the path to achieving SMED can be realized with the commitment to continuous improvement, using lean tools like TPM, and the 5S pillars that are described under sort, set in order, shine, standardize and sustain. It is also imperative to consider that financial inferences accrue in the effort to achieve SMED, and so the plant must be willing to incur costs in the form of human resource training, additional maintenance and time.

With a commitment to continuous improvement, the plant is likely to reap the benefits of incorporating lean tools in its process improvement activities.

FUTURE WORKS

The chapter details critical information about the implementation of lean strategies in the manufacturing setting. The findings present promising results, which can further be extended in future works, not only in manufacturing plants, but other sectors as well. The findings have also shown that one cannot stop at just introducing lean strategies and leaving it at that, instead, regular assessments in the effort to achieve desirable future states have to be considered, indicating the need for developing sustainable action plans. As such, firms should consider using lean strategies to better their process, and action plans must be identified along the way. Future studies should consider evaluating the extent to which organizations are introducing, implementing, assessing, sustaining lean strategies, and the benefits that they are accruing as a result.

REFERENCES

A Lean Journey. (2010). *Quick Changeover*. Retrieved July 23, 2017, from http://www.aleanjourney.com/2010/08/quick-changeover.html

Al-Hassan, K., Chan, J. F. L., & Metcalfe, A. V. (2000). The role of total productive maintenance in business excellence. *Total Quality Management*, *11*(4-6), 596–601. doi:10.1080/09544120050007931

Alukal, G., & Manos, A. (2006). *Lean kaizen: A simplified approach to process improvements*. Milwaukee, WI: American Society of Quality Press.

Bateman, N., & David, A. (2002). Process improvement programmes: A model for assessing sustainability. *International Journal of Operations & Production Management*, *22*(5), 515–526. doi:10.1108/01443570210425156

Bhasin, S. (2011). Performance of organisations treating lean as an ideology. *Business Process Management Journal*, *17*(6), 986–1011. doi:10.1108/14637151111182729

Bialek, R., Duffy, G. L., & Moran, J. W. (2017). *Spaghetti Diagram*. Retrieved July 26, 2017, from http://asq.org/learn-about-quality/process-analysis-tools/overview/spaghetti-diagram.html

Bradbury, J. (2017). *Lean Team*. Retrieved July 27, 2017, from https://www.graphicproducts.com/articles/lean-team/

Carrizo Moreira, A., & Campos Silva Pais, G. (2011). Single minute exchange of die: A case study implementation. *Journal of Technology Management & Innovation*, *6*(1), 129–146. doi:10.4067/S0718-27242011000100011

Collins English Dictionary. (n.d.). Retrieved July 27, 2017, from https://www.collinsdictionary.com/dictionary/english/engineer

Dave, Y., & Sohani, N. (2012). Single Minute Exchange of Dies: Literature Review. *International Journal of Lean Thinking*, *3*(2), 27–37.

Gregory, M. (2012). *Lean Sustainment: Engaging Leadership is the Missing Link*. Retrieved July 26, 2017, from http://www.ipa-involve.com/news/lean-sustainment/

Harris, G., & Donatelli, A. (2005). *Value-stream mapping as a lean management tool*. Huntsville, AL: University of Alabama.

Infinite Innovations Ltd. (2011). *What is Brainstorming?* Retrieved July 27, 2017, from http://www.brainstorming.co.uk/tutorials/whatisbrainstorming.html

Kaizen Institute. (n.d.). *What is Kaizen?* Retrieved from https://ca.kaizen.com/about-us/definition-of-kaizen.html

Lean Manufacturing Tools. (2017). *Development of TPM*. Retrieved July 25, 2017, from http://leanmanufacturingtools.org/430/what-is-tpm/

Lista International Corporation. (2017). *Implementing 5S Workplace Organization Methodology Programs In Manufacturing Facilities*. Retrieved July 25, 2017, from http://www.listaintl.com/resource-center/white-papers-and-case-studies/white-papers/implementing-5s-workplace-organization-methodology-programs-manufact

Manufacturing Terms. (n.d.). *Quick Changeover*. Retrieved July 24, 2017, from http://www.manufacturingterms.com/Quick_changeover.html

Mika, G. (2005). *Kaizen event implementation manual*. Society of Manufacturing Engineers.

Miller, J. (2005). *Introduction to lean and value-stream mapping*. Gemba Research.

Mostafa, S., Dumrak, J., & Soltan, H. (2013). A framework for lean manufacturing implementation. *Production & Manufacturing Research*, *1*(1), 44–64. doi:10.1080/21693277.2013.862159

Parker, J. (2012). *5s and Kaizen for Process Improvement*. Retrieved from http://enfocussolutions.com/5s-and-kaizen-for-process-improvement

Smartware Group. (2016). *Weighing the costs and benefits of TPM implementation*. Retrieved July 25, 2017, from http://www.plantengineering.com/single-article/weighing-the-costs-and-benefits-of-tpm-implementation/b6920afa4d71bb7054417455bd471472.html

Soderquist, K., & Motwani, J. (1999). Quality issues in lean production implementation: A Case Study of a French Automotive Supplier. *Total Quality Management*, *10*(8), 1107–1122. doi:10.1080/0954412997091

Standard, C., & Davis, D. (2000). Lean thinking for competitive advantage. *Automotive Manufacturing and Production*, *12*, 1–3.

Steinlicht, C. L. (2010). *Lean production and the organizational life cycle: A survey of lean tool effectiveness in young and mature organizations* (Order No. 3412450, Capella University). ProQuest Dissertations and Theses, 218.

Stone, K. B. (2012). Four decades of lean: A systematic literature review. *International Journal of Lean Six Sigma*, *3*(2), 112–132. doi:10.1108/20401461211243702

Sun, S. (2011). The strategic role of lean production in SOE's development. *International Journal of Business and Management*, *6*(2), 160–168. doi:10.5539/ijbm.v6n2p160

Tapping, D., Luyster, T., & Shuker, T. (2002). Value stream management. New York, NY: Productivity Press.

Taylor-Powell, E., & Steele, S. (1996). *Collecting evaluation data: Direct observation*. Program Development and Evaluation. Wisconsin: University of Wisconsin-Extension.

Ulutas, B. (2011). An application of SMED Methodology. *World Academy of Science, Engineering and Technology*, *79*, 101.

Upadhye, N., Deshmukh, S. G., & Garg, S. (2010). Lean manufacturing for sustainable development. *Global Business and Management Research*, *2*(1), 125–137.

Velaction Continuous Improvement. (2016). *Changeover*. Retrieved July 24, 2017, from http://www.manufacturingterms.com/Quick_changeover.htmls

Vorne. (2016). *SMED (Single-Minute Exchange of Dies)*. Retrieved July 25, 2017, from http://www.leanproduction.com/smed.html

Weigh Label. (2017). *TPM and Single-Minute Exchange of Die (SMED)*. Retrieved July 24, 2017, from http://weighlabel.com/2014/08/tpm-and-single-minute-exchange-of-die-smed/

Womack, J. P., & Jones, D. T. (1994). From lean production to the lean enterprise. *Harvard Business Review*, *72*(2), 93–103.

Wrye, M. (2012). *The Importance of TPM and SMED*. Retrieved July 24, 2017, from https://beyondlean.wordpress.com/2011/11/21/the-importance-of-tpm-and-smed/

Chapter 7
Dell's Just-in-Time Model in West Africa

ABSTRACT

Dell Computer Corporation founded by Michael Dell started in 1985 and has become the world's leading computer systems company today. Dell's motto is simple: to design, build, and customize products and services to satisfy a range of customer requirements. A few years after introducing its next-day on-site service, which will become the company core of its competitive edge, Dell started to expand internationally with the opening of many subsidiaries. This chapter assesses the feasibility of implementing Dell's model and its impacts on a third-world economies such as Ghana. Key factors such as the Ghana natural and human resources are assessed. The findings show that certain attributes must be in place to facilitate the JIT business model, especially when looking at emerging markets like Ghana with a clue on what it takes to implement the business model application in manufacturing, service, or production settings.

INTRODUCTION

Dell's direct model technology is founded on listening to customers and delivering innovative technology and services to them. Using the direct model technology, Dell sells computers directly to consumers, which gives them a front view of what they want so the company can provide them with the most efficient solution in their capacity. The direct model technology

DOI: 10.4018/978-1-5225-4062-5.ch007

starts and ends with the customers, with the power of direct and a team of talented people, Dell is able to provide customers with superb value, high technology, high quality, customized systems, superior service and support and products and services that are easy to buy and use. (Dell, Inc., 2006)

This is the cornerstone of the direct model technology. According to the website, the model has five tenets:

- Most efficient path to the customer
- Single point of accountability
- Build-to-order
- Low-Cost Leader
- Standard-based Technology.

The model is the most efficient path to the customer because of the direct relationship. There are no intermediaries, which has resulted in less time of delivery and interaction. They directly handle customers to know their specific needs and provide those needs with their talented workforce. Single point of accountability means that Dell makes it easy for the customers by only referencing a single and streamlined way to the right resources. This means that customers are not shuffled into different departments or technical group to answer any questions they asked or fulfill any requirements that need to be fulfilled. As it is, technology can be quite complex and difficult to understand. Dell recognizes this and allows customers not to feel intimidated by the complexities of technology with their single point of accountability. Customers want the purchase to be as painless as possible and to be a customer service experience they would want to have again. The third tenet is built-to-order which means they build exactly what the customer's specifications are. They assess the needs of the customer and they provide them according to their needs. This will lead the customer to more intelligent choices and they can get the best prices for the features they want in a computer which leads to the fourth tenet called a low-cost leader. Dell said, "they focus resources on what matters to the customers". Dell has an outstanding supply chain and manufacturing facilities which provides them with high-quality products at reduced costs. This, in turn, translates to superior value for their customers. The fifth tenet is standards-based technology which means that Dell does extensive research and development to provide customers with standards.

Standards will give customers a wider array of choices which satisfies every specification they need.

The chapter assesses the extent to which success can be achieved when Dell implements its direct model technology in Ghana, and so the project considers understanding the current state of this West African country. The identification from the study is that this developing world country presents potential as a technology hub for West Africa and Africa as a whole, but it will have to address the current loopholes that are in place. For instance, the technological capabilities impairments, poor market opportunity, poor learning environment, low investment from the private sector, heavy taxation imposed by the government on businesses, and volatility in the economy of the country characterized by inflation, among other things, present hindrances to the country's progress. The above disparities indicate that the application of quality management systems in a number of the areas of the country are obsolete or are not working for the sectors. It is high time that the country addresses the above issues among others, and thus makes use of lean six sigma strategies like just-in-time tools in its manufacturing quests, with international partners, in the attempt to foster competitive advantage and significant growth.

The Impact of the Dell's Direct Model Technology in the Development of Ghana

We paid the price of not taking part in the Industrial Revolution ... because we did not have the opportunity to see what was taking place in Europe. Now we see that information and communication technology has become an indispensable tool. This time we should not miss out on this technological revolution. (Allotey, 2000)

Is Ghana ready for Dell's direct model technology? In order to assess the impact of Dell to the economy of Ghana, we have to look into the technical efficiency levels of Ghana to support Dell's direct model technology. Dell, Inc. is operating in several countries already. Their headquarters is in Texas, and they have regional business units in the United States, Canada, and Latin America. Other regional headquarters include "Bracknell, England, for Europe, the Middle East and Africa; Singapore, for Asia-Pacific; and Kawasaki, Japan, for the Japanese market." (Dell, Inc., 2006). The company has several manufacturing facilities in Austin, Texas; Nashville, Tenn.;

Eldorado do Sul, Brazil (Americas); Limerick, Ireland (Europe, Middle East, and Africa); Penang, Malaysia (Asia-Pacific and Japan) and Xiamen, China (China). Dell maintains sales offices in 34 countries around the world and sells its products and services worldwide. Accordingly, Dell has affected the economy of these communities by hiring technical people, skilled personnel and other professionals to form part of the Dell experience. Also, when Dell decides to build a manufacturing facility for its direct model technology, it would have to invest a huge amount of capital and this would mean that jobs will be created for the people in the locality where the facility will be built. Aside from that, Dell would have to procure local contractors and suppliers for the construction. When the facility has been built, the operations of the company would require a local supply chain which Dell is most famous for developing. Dell is able to offer consumers lower-priced products because of their outstanding supply chain management. It is therefore logical that when they do decide to establish their direct model technology in Ghana, they would have to utilize local suppliers to complement imported products that are not available locally. This factor alone would benefit the Ghanaian economy. Just like what they did in their most recent facility in North Carolina. The plant was worth about $100 million 750,000 square-foot plant, which reflects a Dell investment of more than $100 million which "will build Dell's award-winning OptiPlex and Dimension desktop systems for customers in the eastern U. S. Demand for its products has made Dell the leading provider of desktop computers in the U.S. and the world" (Noticias, 2006). In terms of investment on human resource, Dell hired certainly, when Dell opens in Ghana, it would greatly impact its development. In order to deliberate whether Ghana has sufficient infrastructure to support Dell, we would have to research deeply the workings of its economy.

Assessment of Ghana

To assess the technological capability of a country, we need to look at various indictors and not just its technical efficiency. We would also have to study its geography, population, culture, natural resource, production, manufacturing and human resources. Ghana has undergone several structural reforms already and is one of the countries in Africa that "span diversity of per capita incomes and industrial development patterns of the Africa region" (Biggs et al, 1995).

GEOGRAPHY

Ghana is located in West Africa and its capital is Accra. According to Wikipedia, it has a population of a bout 21,029,853. Ghana is the first country in Africa to obtain its independence from the United Kingdom in 1957. Ghana is primarily an agricultural economy which bases its development in agriculture since gifted with a lot of natural resources. Among the countries in Africa, it has a good per capital income but still, Ghana is dependent on financial assistance from donors and wealthier countries. Ghana is located near the equator in the Gulf of Guinea. The Republic of Ghana covers an area of 239,456 square kilometers of which around 29,000 square kilometers is under cultivation and 1,000 under inland water (rivers, lakes, etc.)" (UNDP, 1992) One third of the land is a forest zone and they have an artificial lake called Volta Lake built to generate electricity and provide inland navigation. This lake is an important resource for irrigation and fish farming. Its climate is tropical and its land is mostly covered by "low bush, park-like savanna and grassy plains" (UNDP, 1992). Its eastern coast is warm and comparatively dry while the south-west corner is hot and humid. The north is hot and dry but characterized by rain during July and August. Just like any other tropical country, their seasons are divided into wet and dry only. Does geography play a role in the development of a nation's economy? According to Hausman (2005) in his article entitled "Prisoners of Geography", there is such a thing as the tyranny of geography especially for countries in Africa which as much as they enjoy a tropical climate, they would never enjoy access to markets and new technologies. Ghana is in Africa and has a tropical climate, does this mean that they will never enjoy a better quality of life as compared with other nations? Further, according to the same article, US Treasury Secretary Lawrence said that whatever these poor countries are suffering from, be it "poor soil, erratic climate and vulnerability to infectious disease" should not be put upon them because of their location since "these can be traced simply by the failure of their governments to put in place the right". But as economist suggested, access to markets is crucial for economic growth and countries in Africa are not really positioned in terms of accessibility. This is probably the reason why Hausman offered the proposition that some countries are prisoners of geography. In Ghana's perspective, this may be true. But if Dell chooses Ghana precisely because of its geography in terms of accessibility to most African countries, then Geography, in this instance is not a wall that will hinder them from development.

DEMOGRAPHY

In terms of demography, Ghana is composed of four major tribes and 2% of European descent. 63% are Christians, 21% indigenous belief and 16% are Moslems (UNDP, 1992). Their official language is English. This is a very important factor as English is the business language of the global economy. In Ghana, English is the official and commercial language is taught in all schools. About 25% of their general population is literate in English and the rate is much higher in the younger generation. In the capital Accra, 3 million people live there. According to the US Census, the population in Ghana tripled after the declaration of independence from 6 million to 18 million in 1996, It is expected to rise to 27 million in the year 2020 (Adlakha, 1996). According to UNDP, most people live on the coast and of course, in the major cities such as Accra and Kumasi.

EDUCATION

In the education front, Ghana has relatively "easy access to good education with 12,630 primary schools, 5,450 junior secondary schools, 503 senior secondary schools, 21 training colleges, 18 technical institutions, two diploma-awarding institutions and five universities serving a population of 18 million" (UNESCO, 2004). The Ministry of Education is the department primarily responsible for the education of Ghanaian. Primary and middle school education is tuition-free and is mandatory. They had problems with few teachers than students but this is being addressed by the government because they believe that education would figure highly in their economic development. After primary school, students who pass the Common Entrance Examination will go to the secondary level and entrance to universities are afforded to young people who pass the examination after secondary school graduation. Tertiary education is for six years or more depending on the degree of course. According to the UNESCO report, tertiary education enrollment totals almost 2 million: 1.3 million primary; 107,600 secondary; 489,000 middle; 21,280 technical; 11,300 teacher training; and 5,600 universities. There are five public universities in Ghana and several technical and vocational schools. Technical institutes offer several programs such as "pre-technical courses, general technical courses and craft courses. The pre-technical course, which requires the Common Entrance Examination for admission, takes two years of full-time study to complete. Then students can either take up an apprenticeship or continue their technical education, either in general technical programs or craft courses" while polytechnics offer an additional program in institutional management, catering and domestic subjects leading to the Institutional Management Certificate in Institutional Housekeeping and Catering" (Sedgwik, 2006).

NATURAL RESOURCES

As mentioned earlier in the paper, Ghana is endowed with rich natural resources from "fishing, arable land, forest, sizeable mineral deposits as well as crude oil reserves and natural gas" (IMF, 2000). In fact, several of its lakes and rivers are viable for hydroelectric power plants. Its mineral resources include gold, timber, industrial diamonds, bauxite, and manganese. In terms of water resources, "annual internal renewable water resources: 1,607 cu Per Capita; Sector withdrawals - domestic 35%; Industrial 13%; Agricultural 52%" (Institute for Security Studies, 1998). Natural hazards would be the dry, dusty, harmattan winds which occur from January to March. Aside from that, Ghana has to contend with droughts.

ECONOMY

The primary economic activity in Ghana is agriculture which is mostly small-scale and accounts for 36% of their GDP in 2000. 60% of their workforce is in agriculture while 25% of their GDP is from their main export crop, cocoa. These exports generate about 30-40% of their foreign exchange earnings. Other sources of foreign exchange are gold and timber. In the industrial sector, 25% of their GDP comes from manufacturing, which compared with other Sub-Saharan economy, is "relatively well developed and diverse" (Ghana Government Website, 2006). In the labor market, Ghana's labor force numbered 8.7 million in 1998 and 12 million in 2001. Its industrial sector which is focused on aluminum smelting, sawmills, timber and agricultural processing plants, brewing, cement manufacture, oil refining, textile, electrical, pharmaceuticals, mining, and others. In the service sector which accounts for 32% of the GDP consists of trade and public services. According to the Ghana Stock Exchange, the stock market has improved considerably in the last few years because of the recent efforts to develop tourism. Ghana's government relationship with the World Bank and IMF was considered good since the late 1980s as Ghana has followed their advice and implemented structural programs in the government. There was the Vision 2020 program which guided economic policy and the 1998 – 2000 ESAP program with the IMF. Although Ghana fell short of the performance targets set by these international organizations, the donor community has continued to support Ghana. Of course, debt has spiraled with interest payments allotted by the government reaching to 19% of the national budget. This debt burden is hoped to be lifted with the move for external debt relief in the Sub-Saharan African region for heavily indebted poor countries (World Views, 2006).

Steps Employed in the Implementation of Dell's JIT (Direct Model Technology) Strategy

The first step in JIT method implementation is dubbed as awareness revolution by Phogat (2013). The step requires that the plant and the people therein are prepared for the presence of short lead times, flexibility, low-costs, and high-quality, while paying attention to set up time, people, quality, maintenance, design and layout (Kumar, 2010). After the above awareness, the next step would be to find ways to improve the workplace. This would result in an instance where there is zero lead time, and nil waste as a result of paying attention to absolute involvement of people, collection of process data, visibility, control of inventory, reducing the lot size, enhanced improvement, customer and supplier relations, and flow scheduling, 5s, standard operations, and multi-processing (Kumar, 2010).

The JIT methodology in practice commences after a customer places an order. With the order at hand, a firm goes ahead to manufacture the products that are outlined as per the desires of the customer, indicating the presence of a pull system. A pull system is reactive in nature in the sense that the execution of production is done after a customer places an order. JIT has worked for Dell, and it is in fact considered to be one of the companies that have successfully applied the JIT methodology. In fact, Dell has gone ahead to be a market leader in the area of computer technology because of its dynamic supply chain.

Despite the above information on the implementation and usefulness of JIT, it is worth to note that the methodology has limitations, which take note of; differences in culture, conflict with companies that hold the traditional stock production, culture differences in implementing JIT (applicable in Japan but could not be applicable to other countries, loss of team and individual autonomy, and overall, resistance to change).

Factors That Affect the Development of Dell Direct Model Technology

Dell Computer Corporation has annual revenue of $35.4 billion and considered to be the premier provider of computing products and services. This was supposed to be the result of its direct business model which made "Dell is the leading seller of computer systems worldwide and the number one seller in all customer segments in the United States" (Dell, Inc., 2006). Dell's business strategy is hinged on its direct customer model combined with a highly efficient manufacturing and supply chain management. This has made them provide customers with "superior value; high-quality, relevant technology; customized systems; superior service and support; and products and services that are easy to buy and use" (Dell, Inc., 2006). With their five key tenets, Dell's direct model technology's success is the effect of several factors. One of the critical success factors would be its supply chain management. Dell's suppliers are located in close proximity to its manufacturing facilities. This is crucial to maintaining their direct model, which claims next day on-site delivery. Because of the close proximity of its suppliers, Dell has managed a lean inventory which is only about 5 days' supply (Beals, 2003). Another factor, which affects the development of the model, is their research and development. Dell has enabled to produce products which the customers wanted because of their efficient and up-to-date research and development. Their R&D has innovated a lot of products and this has become one of their competitive edges. Another factor would be their channels of distribution. Dell's direct model technology has been effective in major markets but with smaller markets, they use distributors. Another factor of their successful model is their ability to sense changes in the market. Foresight in planning and product innovation has propelled Dell into the forefront of the computer system industry. The direct model technology would not have been effective for Dell if they do not have an outstanding sales infrastructure to support it. They have direct sellers which account for 50% of their business. These direct sales representatives are the ones who manage key and huge accounts. These are the major businesses and corporations who provide the majority of Dell's profit. Although Dell is famous for its direct selling to customers, individual consumers comprise only a fraction of its business. These individual buyers are respondents to Dell's tri-media ads. Another factor that affects their direct model technology is the inputs of their customers (Schiffelbein, 1999). This is Dell's key strength, which gives them a competitive advantage over the other players in the industry. They receive input from customers regarding their products or potential products and they have the ability to produce a line of products with the customers' inputs in mind.

HUMAN RESOURCES

As of May 2006, Dell provides work for approximately 75,100 employees around the world which gives them a large and diverse corporate organization (MSN Money, 2006). When it opened its Nashville operations, the company had to increase the number of their employees by 1,000. They foresee it rising to about 3,500 when it opens its call center in Tennessee. Its total investment in Tennessee is $250 million with its three manufacturing facilities. The total percentage of employees in the US is only 48.3%, the rest of the 51.7% are working outside of the United States. In terms of a benefits package, Dell allows employees to choose from a "selection of health, dental and vision plans or build their own and they also offer flexible spending accounts and life insurance coverage" (New York Job Source, 2006). In terms of retirement, employees can invest 25% of their annual pre-tax salaries and the company will match 3% of this. Each employee is eligible to participate in the Dell 401. Aside from this, Dell offers an Employee Stock Purchase Plan with a 15% employee discount from the listed price. Payment of stock purchase can be done through salary deduction. Other benefits Dell provide their employees according to the New York Job Source are paid vacation based on years of service, paid holidays, including an extended week for Christmas/New Years, Personal Business Allowance (PBA) Dell offers paid time off for personal reasons such as; illness, medical appointments, closing on a home, and childcare obligations. leave of absence where Dell provides eligible employees with up to 12 weeks of unpaid, job-protected leave in a rolling 12-month period for certain qualifying conditions in accordance with the federal Family and Medical Leave Act of 1993 (FMLA). Dell also provides Military Leave and, under certain circumstances, Company Discretionary Leave, "Friends For Hire" or the Employee Referral Program, adoption assistance, discounts on Dell products and an employee fitness center.

Dell Computer made a "conscious effort to hire employees who had open, questioning minds and were always ready to learn and try something new".

Potential employees and job candidates go through a screening process. Dell looks for people who are not just results-oriented, self-reliant, and intelligent but people who are flexible and not opposed to change and can look at a problem with a different solution. They expect employees to have different perspectives and can come up with alternative and innovative solutions and ideas. Employees are expected to grow and develop with the company. As such, they should have the capacity to change with the company and not be hindered by structures. They should be able to look for better ways to do things and question how "things were done and offer suggestions for improvement".

CULTURE

According to its corporate website, Dell's culture is hinged on a passion for winning in everything they do. It stated that "that the effort of reaching for more and better is not negotiable. Continuous improvement is part of our DNA and drives our efforts to be a great company and a great place to work. In short, cultivate a Winning Culture. A Winning Culture requires talented individuals working as a team with a commitment to common values and a collective determination to execute on a shared vision". As such, they continue to develop this winning culture by the "development of leadership skills, encouraging personal accountability, maintenance of a higher standard of ethics and integrity and having a workforce with a diverse background, skills and potential". Accordingly, it said that the soul of Dell is the statement of their values, beliefs, and aspirations. They believe that the soul of their company is all its employees who "demonstrate their commitment to the customers, the Dell teams, direct relationships, global citizenship and winning" (Dell, Inc., 2006). But this was not the culture of Dell before. In the 1990s, CEO Kevin Rollins lamented that the corporate culture of Dell was "culture of stock price, a culture of financial performance, and a culture of 'what's in it for me" (Hamilton, 2004). Rollins and his management team did a soul searching and they felt that Dell should have a change of culture as soon as possible. Their transformation though not so easy was not so difficult to do because of Dell is a prototypical flat organization. From the factory floor to corporate communications, decisions are made quickly and without the burden of superfluous hierarchy. Management's accessibility has made communications flow smoothly and quickly that the change in culture was not really hard to implement. As one of the employees stated "Dell is a meritocracy: if you work hard and do a good job, you get rewarded for it. It also says that just getting the job done is not as important as how you do it, and the quality of the teamwork you do it with" (Hamilton, 2004). Employees believed that Dell management has supported their efforts which help them achieve work/life balance and they are able to see the relationship between their work and the company objectives.

Ghana's Advantages and Limitations to Develop This Technology Advantages

Ghana has a relatively good international reputation with respect to political stability and macroeconomic reforms. In terms of private sector development, however, there are mixed views on Ghana. In 2001, Ghana's government has proclaimed the year as the "Golden Age of Business" and Ghana's aim to become competitive in the business sector has provided some reforms in their macroeconomics. The government has emphasized the need for private sector development and attracting foreign direct investment of which Dell's investment could be categorized. After the 196 elections until 2000 the economy stabilized somewhat. Ghana's economy was growing at around 4% per annum and gross capital formation stayed above 20% of GDP although FDI declined (Wolf, 2003). The Ghanaian economy has grown from 1990 to 346 GDP per capita to 2000 with 413 GDP per capita. This is an average annual growth rate of 3.70%. In terms of labor, Ghana offers cheap labor, which Dell can capitalize on. According to the report compiled by the World Bank on wage and productivity gaps in Ghana, levels of wages in the country is not too high to compete internationally. Also, there is evidence that human capital through training and education has increased because of the economic policies which are being implemented aimed at increasing productivity in Ghanaian companies.

LIMITATIONS

The major obstacle in private sector development in Ghana is its macroeconomic environment. According to the study done by the Center for Development Research, the high and volatile inflation rates were prevalent in Ghana and it has affected the capital markets which lead to inefficient capital resource allocation. Low investment by the private sector in Ghana is caused by low savings levels and other risk factors. In terms of market size, Ghana's market size is small, but at the micro level, this would only affect local firms. If Dell will invest in Ghana, the infrastructure that they would have to develop in the country would have to commensurate the market. If Dell's objective is to make Ghana the hub of the African region, it would have to increase their infrastructure size to their target market outside of Ghana. Other limitations that can be seen in Ghana is its taxation which would have an effect on manufacturing, investment incentives by the government which is very important for the non-industrial sector, it's market potential which is limited. In terms of structural adjustments that would support technology such as Dell's, African governments' response has been sluggish. Technological capabilities are touted to be the core of economic reforms. If it wants to be competitive in terms of technology, Ghana would have to narrow the gap between their technological capabilities with other parts of the world. Another gap that would have to be eliminated is the learning environment in the region. It appears that Africa as a region has a weak learning environment although big strides have been made in Ghana.

CONCLUSION

The successful democratic change of government in 2001 makes Ghana an investor- friendly country. Inflation became under control and there were various economic reforms that were implemented which make the country's future for private sector development promising. In order for Ghana to become more attractive to investors such as Dell, it should further improve its social, educational and health infrastructure which is critical in shaping the investment appeal of a country. In industries such as IT, transportation and communications factors are critical, and this is particularly vital to Dell whose supply chain management is a key factor in its direct model technology. Logistics infrastructure for Dell is very important as this is where their competitive edge is based on. If Ghana's transport and communication infrastructure improve, it would be a good site for the development of Dell's model. Ghana should further develop all sectors of the economy in order to attract private sector development in the country. Dell's globalization strategy means that they would have to look at the potential expansion to other countries and Ghana is in a place that can provide Dell with a specific location that would help the company realize its business objectives. If all the necessary support is in place in Ghana, it would be to Dell's advantage to invest in the development of their direct technology in the country. This would benefit both Ghana and Dell. Ghana's economy will certainly improve as Dell would have to source out local suppliers and will open many jobs for its workforce. Dell would have a strategically placed facility that could handle its African operations. Dell's international expansion would increase their market share, revenues, profit, and buyer awareness. However, the Dell Direct Model Technology might not work everywhere. Ghana is an emerging country and in terms of technology, a distributorship would work in the country better than the full-scale direct model approach. Certainly, Ghana would offer a market for computer systems but it is not a big market that would necessitate the whole direct assault. Dell would have to judge that their products should be a commodity where the demand is already in place and it critical that Ghana's support system can handle the value on-time delivery of Dell. In the keynote speech of the minister of education of Ghana, Prof. Christopher Ameyaw-Akumfi aptly said that "Ghana must create the conditions for African science and technology to flourish" (African Youth Foundation, 2006). It is advantageous for Dell to stake a claim in the computer market in Ghana since it is indeed a flourishing market but in terms of how big the investment should be, Dell would have to take into consideration various factors of the Ghanaian economy to ensure that whatever type of investment they would make, it would be successful. It is said that technology would help save Africa particularly Ghana as it is a force of equity and Dell would have an important part of saving Ghana if it invests in it.

FUTURE WORKS

It is worth to note that sharing the findings in this paper with critical stakeholders in Ghana (government and the private) to foster their understanding of where the country is at as far as the gaps indicted above, and their implications are pertinent. Future works should look into assessing the progress that the country has made as far as the attributes that foster LSS such as technology expansion and market growth (technology products), technological efficiency, lowered business taxation rates, private sector investment growth, and growth in the learning environment. From the above assessment, other studies can be done to ascertain whether Ghana is ready to be the technology hub of Africa, by attracting the investment and success of multinationals like Dell, which can then apply its direct mode of production (JIT) thus securing a competitive edge in the West African market.

REFERENCES

Adei, S. (1987). *Technology Transfer and Nationalization in Ghana*. Ottawa, Canada: IDRC.

Adlakha, A. (1996). *Population Trends: Ghana*. Bureau of Census, United States Department of Commerce.

African Youth Foundation. (2006). *Technology and Human Development in Africa*. Retrieved 27 October 2006 from http://www.ayf.de/symposium_english_2002_part3.php

Allotey, F. (2000). *Ministry of Communications of Ghana: Plan for National Information and Communications Infrastructure, 2000–2005*. Academic Press.

Arthur, L. W. (1989). *Some Aspects of Economic Development*. Accra: Ghana Publishing Corp.

Beals, T. (2003). *Dell Case Study*. Retrieved 27 October 2006 from http://www.creative-media-services.com/CMS_Dell_Computers_Case_Study.pdf

Biggs, T., Shah, M., & Srivastava, P. (1995). *Technological Capabilities and Learning in African Enterprises*. Washington, DC: The World Bank.

Dalitso, K. (2000). *The Policy Environment for Promoting Small and Medium-Sized Enterprises in Ghana and Malawi*. Manchester, UK: Institute for Development Policy and Management.

Data Quest. (2000). *Dell's Direct Model*. Retrieved 23 October 2006 from http://www.dqindia.com/content/top_stories/100110402.asp

Dell, Inc. (2006). *Company Overview and History*. Retrieved 24 October 2006 from http://www.dell.com/

Dell, M., & Fredman, C. (1999). *Direct from Dell*. Collins.

Dorte, V. (1999). *Wage and productivity Gaps: Evidence from Ghana*. Washington, DC: The World Bank.

Ghana Government Website. (2006). *Overview*. Retrieved 26 October 2006 from http://www.ghana.com/

Haggarty, L. (2003). *Telecommunication Reform in Ghana*. Washington, DC: The World Bank.

Hamilton, A. B. (2004). How Dell Got Soul. *CNet News*. Retrieved 27 October 2006 from http://news.com.com/How%20Dell%20got%20soul/2030-1069_3-5366494.html

Hausman, R. (2005). *Prisoners of Geography*. Carnegie Endowment for National Peace.

International Development Association. (2001). *Private Sector Development Project: Republic of Ghana, Implementation Completion Report*. Washington, DC: The World Bank.

International Monetary Fund (IMF). (2000). *Ghana: Economic Development in a Democratic Environment*. Washington, DC: IMF.

Institute for Security Studies. (1998). *Ghana's Natural Resources*. Retrieved 25 October 2006 from http://www.iss.co.za/AF/profiles/Ghana/NatRes.html

Kumar, V. (2010). JIT based quality management: Concepts and implications in Indian context. *International Journal of Engineering Science and Technology*, *2*(1), 40–50.

Kusi, N. K. (1991). *The Management and Performance of the Ghanaian Economy, 1960-82*. Armidale: University of New England, Faculty of Economic Studies.

MSN Money. (2006). *Dell Company Report*. Retrieved 27 October 2006 from http://moneycentral.msn.com/investor/research/profile

New York Job Source. (2006). *Dell*. Retrieved 27 October 2006 from http://nyjobsource.com/dell.html

Noticias. (2006). *Dell Opens North Carolina Manufacturing Facility*. Retrieved 24 October 2006 from http://www.noticias.info/asp/aspComunicados.asp?nid=106167

Paul, S. (1990). *Assessment of the Private Sector: A Case Study and is Methodological Implications*. Washington, DC: The World Bank.

Phogat, S. (2013). Introduction to JIT: A review. *International Journal of Latest Research in Science and Technology*, *2*(6), 97–101.

Schiffelbein, M. (1999). *Dell Computers*. Topeka, KS: Emporia University.

Sedgwik, R. (2006). Education in Ghana. *World Education News and Reviews*. Retrieved 24 October 2006 from http://www.wes.org/ewenr/00march/practical.htm

Yankey, G. S. (1987). *International Patents and Technology Transfer to Less Developed Countries: The Case of Ghana and Nigeria*. Aldershot, UK: Avebury.

World Bank. (1991). *Small Enterprises in Ghana: Responses to Adjustment*. Washington, DC: The World Bank.

Tsikata, Y. (2001). *Owning Economic Reforms: A Comparative Study of Ghana and Tanzania*. Helsinki: UNU/WIDER.

United Nations Development Programme (UNDP). (1992). *Living Conditions in Ghana*. Accra: UNDP.

UNESCO. (2004). *Institute for Statistics & National Technical Committee on Educational Statistics Capacity Building:* Educational Statistics Capacity Building Programme in Support of the Republic of Ghana – Diagnostic Report. UNESCO.

Wolf, S. (2003). *Private Sector Development and Competitiveness in Ghana*. Bonn: ZEF Center for Development and Research.

World Views. (2006). Africa's Debt Burden - Who Suffers? *Africa World Press Guide*. Retrieved 27 October 2006 from http://worldviews.igc.org/awpguide/debt.html

Conclusion

The book emphasizes the utilization of Lean Tools, Lean Six Sigma principles, and Quality Process Improvements in a variety of applications. Principal Lean Tools such as Value Stream Mapping (VSM), Flow Charts, and Continuous Improvement Charts; as well as Quality Improvement problem-solving techniques such as Six Sigma DMAIC, Attribute Control Charts, PDCA, Pareto Charts, Ishakawa (Fishbone) Diagrams, Histograms, Scatter Diagrams, ISO 9000 standards have been utilized for performance assessment in manufacturing and warehousing. The text also highlights the use of these Lean / Quality tools to other transactional process settings, including academic/commercial facilities, medical facilities, civil engineering, and the emerging computer, with Dell Computer Corporation, as a case study.

In Chapter 1, the six Sigma DMAIC (Define, Measure, Analyze, Improve, and Control) methodology was utilized for analyzing rework performance of bearing demonstrated that bearings which fail due to low band noise, produce a low final yield after being reworked. On the other hand, bearings that fail due to high band noise, had a much higher final yield after being reworked. Attribute Control Charts were extensively utilized in the study in Chapter 1. With the aid of MINITAB 17, Proportion-defective (p) charts were generated. The p chart of low band reworked inners showed an average p_{bar} of 0.611, indicating that 61% of low band reworked inner bearings were nonconforming. The inclusion of Lean Six Sigma approaches enabled appropriate recommendations to be made to the manufacturing facility on how to approach their low band and high band scrap in the future. An additional recommendation was for the manufacturing facility to send bearings that are rejected for high band to be reworked on the hones and sent back through assembly. In the study Value Stream Maps (VSM's) proved useful in assessing the current state of the bearings rework process from which a future state VSM was established to improve efficiency in the rework process.

Since quality assessment are determined based on statistical analysis, Chapter 2 emphasized the use of T-test and Analysis of Variance (ANOVA) as determinants of a physical variable, i.e., the luminosity of lighting on an academic facility. Starting with a Pareto Analysis on the causes to determine to see what the buildings are making up the 80% of the lighting of the campus, the study yielded a P-value of 0.000, indicating that the light intensity of on the facility can be measured up to the national standard found in the literature. This result was backed by a one-way ANOVA conducted on the luminosity of all buildings on the facility whereby the ANOVA test showed a significant difference among the light intensity between all buildings on campus. Regression Analysis was also used for the calculated light intensity for every location on the facility. The use of these statistical tools aided to prioritize in making recommendations for modification, enhancement or addition of exterior lighting for the facility for the next 20 years.

Chapter 3 highlights the use of Lean principles to improve the design of a virtual warehouse learning efforts. The system described in the Chapter is a useful learning tool for students to explore a warehouse virtually, by using an avatar to learn basic warehousing operations. It is vital to appreciate the role that simulations play in learning as far as changing how knowledge is imparted is concerned. Online simulations can be used to reinforce learning, where students can go back to the real world and apply what they have learned. This is where the future of education is headed and there is no better way to foster online learning using simulations that have been designed and improved with the application of the concept of lean Six Sigma.

Chapter 4 dealt with the use of Lean and statistical Quality tools to reduce waste at medical facilities. The study emphasized the notion that by adapting and utilizing lean tools such as a modified Plan, Do, Study, Act (PDCA Cycle), with emphasis on an additional step, "Identify" (for the purpose of clearly identifying root causes), and statistical tools such as scatter diagrams, histograms, Pareto charts, fishbone diagrams and control charts, waste can be identified and reduced or eliminated. This leads to increased satisfaction, efficiency and creates competitive advantage. The study also successfully demonstrated that the quality philosophies and tools such as Lean and Six Sigma SPC can be applied to healthcare and other transactional process settings, aside the traditionally established lean/quality domain of the manufacturing sector.

Chapter 5 extended that areas of application for Lean concepts and Total Quality Management (TQM) tools to the realization for the need for ISO

9000 standards in the profession of Civil Engineering. The study involved researching information concerning the operations of civil engineering consulting firms and their overall management procedures to obtain an understanding of the basic nature of their business culture. Ten firms were selected at random from a geographic area covering the central United States representing a wide range of companies by size. Based on the data collected, the civil engineering consulting firms that recognize and are familiar with the management processes TQM, ISO 9000 and Lean are the four largest firms, with the largest and only international firm having adopted one of the processes, ISO 9000. While this is a broad conclusion based on the small sample selected for review in this study, it appears analogous to an economy of scale being the relevancy of firm size related to degree of complexity of management style for the organizations studied.

The purpose of a lean system was explained in chapter 6 as to increase the throughput of a company to meet customers' product demands and to provide them with what they want, when they want, and at the same time meets the customers' quality expectations. The results of the study presented in the chapter showed that the reliance on systematic strategies using lean tools and principles, and the emphasis on training are the key element to a successful lean deployment in a global supply chain environment. It was also noted that a positive impact is felt on the work area, especially with the presence of team work as shown in the organizational layout of the plant and the crucial role it played in the success of the lean effort. The changes that have been evidenced by the use of the tools identified earlier confirm that the path to achieving SMED can be realized with the commitment to continuous improvement, using lean tools like TPM, and the 5S pillars that are described under sort, set in order, shine, standardize and sustain. It was also imperative to consider that financial inferences accrue in the effort to achieve SMED, and so the plant must be willing to incur costs in the form of human resource training, additional maintenance and time. With a commitment to continuous improvement, the plant is likely to reap the benefits of incorporating lean tools in its process improvement activities.

Finally, Chapter 7, extended the reach of the theme to more geographic settings with a study of Dell Computer Corporation's introduction of its next day on-site service to the developing world's economy, particularly the West African country of Ghana, following Dell's efforts to expand internationally with the opening of many subsidiaries. In industries such as the computer industry, transportation and communications are critical to the supply chain management and the company's their direct model technology. Logistics infrastructure for Dell is very important as this is where their competitive edge is based on. The study showed that Dell's globalization strategy means that they would have to look at the potential expansion to other countries, such as Ghana. If all the necessary support is in place, it would be to Dell's advantage to invest on the development of their direct technology in the developing world.

Related Readings

To continue IGI Global's long-standing tradition of advancing innovation through emerging research, please find below a compiled list of recommended IGI Global book chapters and journal articles in the areas of risk assessment, project management, and portfolio management. These related readings will provide additional information and guidance to further enrich your knowledge and assist you with your own research.

Aboutalebi, R. (2016). Strategies for Effective Worldwide Supply Chains. In B. Christiansen (Ed.), *Handbook of Research on Global Supply Chain Management* (pp. 1–14). Hershey, PA: IGI Global. doi:10.4018/978-1-4666-9639-6.ch001

Afonso, P., & Jiménez, V. J. (2016). Costing Systems for Decision Making under Uncertainty Using Probabilistic Models. In D. Jakóbczak (Ed.), *Analyzing Risk through Probabilistic Modeling in Operations Research* (pp. 221–245). Hershey, PA: IGI Global. doi:10.4018/978-1-4666-9458-3.ch010

Agarwal, R., & Virine, L. (2017). Integration of Project Risk Management (PRM) into Enterprise Risk Management (ERM). In Y. Raydugin (Ed.), *Handbook of Research on Leveraging Risk and Uncertainties for Effective Project Management* (pp. 294–317). Hershey, PA: IGI Global. doi:10.4018/978-1-5225-1790-0.ch014

Agarwal, R., & Virine, L. (2017). Monte Carlo Project Risk Analysis. In Y. Raydugin (Ed.), *Handbook of Research on Leveraging Risk and Uncertainties for Effective Project Management* (pp. 109–129). Hershey, PA: IGI Global. doi:10.4018/978-1-5225-1790-0.ch005

Agostino, D., Arnaboldi, M., & Azzone, G. (2016). Performance Measurement in Public Networks: Developing a PMS for Network Actors and Network Managers. In A. Ferreira, G. Azevedo, J. Oliveira, & R. Marques (Eds.), *Global Perspectives on Risk Management and Accounting in the Public Sector* (pp. 298–319). Hershey, PA: IGI Global. doi:10.4018/978-1-4666-9803-1.ch015

Aiello, L., & Gatti, M. (2017). Project Portfolio Management and Organization: An Integrated and Circular Model. In L. Romano (Ed.), *Project Portfolio Management Strategies for Effective Organizational Operations* (pp. 288–309). Hershey, PA: IGI Global. doi:10.4018/978-1-5225-2151-8.ch012

Alfirević, N., Šešić, M. D., & Čačija, L. N. (2017). The Fundraising Patterns and the Management Professionalization in Independent Arts Organizations in SEE. In V. Potocan, M. Üngan, & Z. Nedelko (Eds.), *Handbook of Research on Managerial Solutions in Non-Profit Organizations* (pp. 400–420). Hershey, PA: IGI Global. doi:10.4018/978-1-5225-0731-4.ch019

Amaral, A. M., & Araújo, M. (2017). Project Portfolio Selection Using a D.E.A. Approach. In L. Romano (Ed.), *Project Portfolio Management Strategies for Effective Organizational Operations* (pp. 220–244). Hershey, PA: IGI Global. doi:10.4018/978-1-5225-2151-8.ch009

Aparo von Flüe, A. (2017). Are You Pondering What I Am Pondering?: Eccentric Consideration on Strategic Management. In L. Romano (Ed.), *Project Portfolio Management Strategies for Effective Organizational Operations* (pp. 81–118). Hershey, PA: IGI Global. doi:10.4018/978-1-5225-2151-8.ch004

Araya-Leandro, C., Caba-Pérez, M. D., & López-Hernandez, A. M. (2016). Modernization of Governmental Accounting Systems: Situation in the Central American Region. In A. Ferreira, G. Azevedo, J. Oliveira, & R. Marques (Eds.), *Global Perspectives on Risk Management and Accounting in the Public Sector* (pp. 90–107). Hershey, PA: IGI Global. doi:10.4018/978-1-4666-9803-1.ch005

Archibald, R. D. (2017). Inter-Relationships between an Enterprise's Strategic Management Process and Its Program/Project Portfolio Management Process. In L. Romano (Ed.), *Project Portfolio Management Strategies for Effective Organizational Operations* (pp. 39–60). Hershey, PA: IGI Global. doi:10.4018/978-1-5225-2151-8.ch002

Arsenijević, O. M., Orčić, D., & Kastratović, E. (2017). Development of an Optimization Tool for Intangibles in SMEs: A Case Study from Serbia with a Pilot Research in the Prestige by Milka Company. In M. Vemić (Ed.), *Optimal Management Strategies in Small and Medium Enterprises* (pp. 320–347). Hershey, PA: IGI Global. doi:10.4018/978-1-5225-1949-2.ch015

Aydin, E. (2017). "Problems and Suggestions": Non-Governmental Organisations of Sexual Orientation Minorities in the Context of Turkey and United Kingdom. In V. Potocan, M. Ünğan, & Z. Nedelko (Eds.), *Handbook of Research on Managerial Solutions in Non-Profit Organizations* (pp. 232–252). Hershey, PA: IGI Global. doi:10.4018/978-1-5225-0731-4.ch011

Ayyıldız, H., Akmermer, B., & Akyüz, A. M. (2017). Marketing Approach for Non-Profit Organizations. In V. Potocan, M. Ünğan, & Z. Nedelko (Eds.), *Handbook of Research on Managerial Solutions in Non-Profit Organizations* (pp. 156–181). Hershey, PA: IGI Global. doi:10.4018/978-1-5225-0731-4.ch008

Balakrishna, P., Borener, S. S., Crook, I., Durston, A., & Robinson, M. J. (2016). Applying Probabilistic Risk Assessment to Safety Risk Analysis in Aviation. In D. Jakóbczak (Ed.), *Analyzing Risk through Probabilistic Modeling in Operations Research* (pp. 321–343). Hershey, PA: IGI Global. doi:10.4018/978-1-4666-9458-3.ch013

Bali, S., & Uslu, Y. D. (2017). Enterprise Risk Management in Non-Profit Organization. In V. Potocan, M. Ünğan, & Z. Nedelko (Eds.), *Handbook of Research on Managerial Solutions in Non-Profit Organizations* (pp. 75–93). Hershey, PA: IGI Global. doi:10.4018/978-1-5225-0731-4.ch004

Baporikar, N. (2016). Business Process Management. *International Journal of Productivity Management and Assessment Technologies*, *4*(2), 49–62. doi:10.4018/IJPMAT.2016070104

Brzozowski, M., & Ferster, I. (2017). Educational Management Leadership: High School Principal's Management Style and Parental Involvement in School Management in Israel. In V. Potocan, M. Ünğan, & Z. Nedelko (Eds.), *Handbook of Research on Managerial Solutions in Non-Profit Organizations* (pp. 55–74). Hershey, PA: IGI Global. doi:10.4018/978-1-5225-0731-4.ch003

Brzozowski, M., & Keydar, E. J. (2017). Approaches towards Measuring Success in Public Elementary Schools. In V. Potocan, M. Ünğan, & Z. Nedelko (Eds.), *Handbook of Research on Managerial Solutions in Non-Profit Organizations* (pp. 362–377). Hershey, PA: IGI Global. doi:10.4018/978-1-5225-0731-4.ch017

Bucero, A. (2017). Linking Organization's Strategy and Strategic Planning with Portfolio Management. In L. Romano (Ed.), *Project Portfolio Management Strategies for Effective Organizational Operations* (pp. 61–80). Hershey, PA: IGI Global. doi:10.4018/978-1-5225-2151-8.ch003

Cakar, U., & Kim, H. (2017). Experiences of Non-Profit Organizations in Korea: Development Challenges in an Emerging Economy. In V. Potocan, M. Ünğan, & Z. Nedelko (Eds.), *Handbook of Research on Managerial Solutions in Non-Profit Organizations* (pp. 499–528). Hershey, PA: IGI Global. doi:10.4018/978-1-5225-0731-4.ch023

Carini, C., & Teodori, C. (2016). Potential Uses and Usefulness of Italian Local Government Consolidated Financial Reporting: The Case of the Town Council of Brescia. In A. Ferreira, G. Azevedo, J. Oliveira, & R. Marques (Eds.), *Global Perspectives on Risk Management and Accounting in the Public Sector* (pp. 68–89). Hershey, PA: IGI Global. doi:10.4018/978-1-4666-9803-1.ch004

Caron, F. (2016). A Bayesian Approach to Project Control. In D. Jakóbczak (Ed.), *Analyzing Risk through Probabilistic Modeling in Operations Research* (pp. 125–137). Hershey, PA: IGI Global. doi:10.4018/978-1-4666-9458-3.ch005

Castagnoli, E., De Donno, M., Favero, G., & Modesti, P. (2016). A Different Way to Look at Random Variables: Do Decision Makers Look at Functions or at Sets? In D. Jakóbczak (Ed.), *Analyzing Risk through Probabilistic Modeling in Operations Research* (pp. 179–199). Hershey, PA: IGI Global. doi:10.4018/978-1-4666-9458-3.ch008

Castagnoli, E., Favero, G., & Modesti, P. (2016). Price Systems for Random Amounts: A Unified Approach. In D. Jakóbczak (Ed.), *Analyzing Risk through Probabilistic Modeling in Operations Research* (pp. 354–379). Hershey, PA: IGI Global. doi:10.4018/978-1-4666-9458-3.ch015

Chandan, H. C. (2016). Implementation of Green Supply Chain Management in a Globalized Economy. In B. Christiansen (Ed.), *Handbook of Research on Global Supply Chain Management* (pp. 402–418). Hershey, PA: IGI Global. doi:10.4018/978-1-4666-9639-6.ch023

Chapman, R. J. (2017). Complexity Theory and System Dynamics for Project Risk Management. In Y. Raydugin (Ed.), *Handbook of Research on Leveraging Risk and Uncertainties for Effective Project Management* (pp. 152–176). Hershey, PA: IGI Global. doi:10.4018/978-1-5225-1790-0.ch008

Chen, E. J. (2016). Simulation Output Analysis and Risk Management. In D. Jakóbczak (Ed.), *Analyzing Risk through Probabilistic Modeling in Operations Research* (pp. 200–220). Hershey, PA: IGI Global. doi:10.4018/978-1-4666-9458-3.ch009

Chen, Y., Davutyan, N., & Ersoy, İ. (2016). A "Flying High, Landing Soft" Platform for Supplier Diversity. In B. Christiansen (Ed.), *Handbook of Research on Global Supply Chain Management* (pp. 72–84). Hershey, PA: IGI Global. doi:10.4018/978-1-4666-9639-6.ch004

Choptiany, J. M. (2017). A Holistic Multi-Criteria Decision and Risk Analysis Method for CCS and GHG Mitigation Projects. In Y. Raydugin (Ed.), *Handbook of Research on Leveraging Risk and Uncertainties for Effective Project Management* (pp. 384–405). Hershey, PA: IGI Global. doi:10.4018/978-1-5225-1790-0.ch018

Coste, A. I., & Tiron-Tudor, A. (2016). Performance Measurement Perceptions in Romanian Higher Education. In A. Ferreira, G. Azevedo, J. Oliveira, & R. Marques (Eds.), *Global Perspectives on Risk Management and Accounting in the Public Sector* (pp. 346–365). Hershey, PA: IGI Global. doi:10.4018/978-1-4666-9803-1.ch017

Çotur, Ö. K., & Öztürkoğlu, Y. (2016). Partnerships in Supply Chain Management. In B. Christiansen (Ed.), *Handbook of Research on Global Supply Chain Management* (pp. 161–185). Hershey, PA: IGI Global. doi:10.4018/978-1-4666-9639-6.ch010

Couto, M. C., & Ferreira, A. D. (2016). The Importance of Accountability Practices in the Public Sector: Literature Review. In A. Ferreira, G. Azevedo, J. Oliveira, & R. Marques (Eds.), *Global Perspectives on Risk Management and Accounting in the Public Sector* (pp. 188–201). Hershey, PA: IGI Global. doi:10.4018/978-1-4666-9803-1.ch010

Cropley, C. H. (2017). The Case for Truly Integrated Cost and Schedule Risk Analysis. In Y. Raydugin (Ed.), *Handbook of Research on Leveraging Risk and Uncertainties for Effective Project Management* (pp. 76–108). Hershey, PA: IGI Global. doi:10.4018/978-1-5225-1790-0.ch004

Csillag, S., Balázs, E., Kocsis, M., Udvarhelyi, T., & Vago, I. (2017). Eight-Point Project: Action Research, as a Development Tool for Non-Profit Organizations. In V. Potocan, M. Üngan, & Z. Nedelko (Eds.), *Handbook of Research on Managerial Solutions in Non-Profit Organizations* (pp. 483–498). Hershey, PA: IGI Global. doi:10.4018/978-1-5225-0731-4.ch022

Cuadrado-Ballesteros, B., García-Sánchez, I. M., & Martínez-Ferrero, J. (2016). Commercialization of Local Public Services. In A. Ferreira, G. Azevedo, J. Oliveira, & R. Marques (Eds.), *Global Perspectives on Risk Management and Accounting in the Public Sector* (pp. 132–150). Hershey, PA: IGI Global. doi:10.4018/978-1-4666-9803-1.ch007

Cuadrado-Ballesteros, B., Mordán, N., & Frías-Aceituno, J. V. (2016). Transparency as a Determinant of Local Financial Condition. In A. Ferreira, G. Azevedo, J. Oliveira, & R. Marques (Eds.), *Global Perspectives on Risk Management and Accounting in the Public Sector* (pp. 202–225). Hershey, PA: IGI Global. doi:10.4018/978-1-4666-9803-1.ch011

Cunha, A., Ferreira, A. D., & Fernandes, M. J. (2016). The Influence of Accounting Information in the Re-Election of the Mayors in Portugal. In A. Ferreira, G. Azevedo, J. Oliveira, & R. Marques (Eds.), *Global Perspectives on Risk Management and Accounting in the Public Sector* (pp. 108–131). Hershey, PA: IGI Global. doi:10.4018/978-1-4666-9803-1.ch006

Ćwiklicki, M. (2017). Applying Balanced Scorecard in Non-Government Organizations. In V. Potocan, M. Üngan, & Z. Nedelko (Eds.), *Handbook of Research on Managerial Solutions in Non-Profit Organizations* (pp. 134–155). Hershey, PA: IGI Global. doi:10.4018/978-1-5225-0731-4.ch007

Damnjanović, A. M. (2017). Knowledge Management Optimization through IT and E-Business Utilization: A Qualitative Study on Serbian SMEs. In M. Vemić (Ed.), *Optimal Management Strategies in Small and Medium Enterprises* (pp. 249–267). Hershey, PA: IGI Global. doi:10.4018/978-1-5225-1949-2.ch012

Daneshpour, H. (2017). Integrating Sustainable Development into Project Portfolio Management through Application of Open Innovation. In M. Vemić (Ed.), *Optimal Management Strategies in Small and Medium Enterprises* (pp. 370–387). Hershey, PA: IGI Global. doi:10.4018/978-1-5225-1949-2.ch017

del Mar Gálvez-Rodríguez, M., Haro-de-Rosario, A., & Caba-Pérez, M. D. (2017). The Relation between Contingency Factors and the Efficiency of NPOs. In V. Potocan, M. Ünğan, & Z. Nedelko (Eds.), *Handbook of Research on Managerial Solutions in Non-Profit Organizations* (pp. 321–342). Hershey, PA: IGI Global. doi:10.4018/978-1-5225-0731-4.ch015

Dieguez, T., Conceição, O., & Fernandes, Â. (2017). The Challenges of Managing the Private Institutions of Social Solidarity: Experiences from Portugal. In V. Potocan, M. Ünğan, & Z. Nedelko (Eds.), *Handbook of Research on Managerial Solutions in Non-Profit Organizations* (pp. 23–54). Hershey, PA: IGI Global. doi:10.4018/978-1-5225-0731-4.ch002

Duong, L. N., Wood, L. C., & Wang, X. (2016). Review of RFID Applications in Perishable Inventory Management. In B. Christiansen (Ed.), *Handbook of Research on Global Supply Chain Management* (pp. 139–146). Hershey, PA: IGI Global. doi:10.4018/978-1-4666-9639-6.ch008

Emel, G. G., & Petriçli, G. (2016). A Staged Supplier Pre-Evaluation Model: To Determine Risky, Potential and Preferred Suppliers. In B. Christiansen (Ed.), *Handbook of Research on Global Supply Chain Management* (pp. 419–451). Hershey, PA: IGI Global. doi:10.4018/978-1-4666-9639-6.ch024

Ercegovac, I. R., Alfirević, N., & Koludrović, M. (2017). School Principals' Communication and Co-Operation Assessment: The Croatian Experience. In V. Potocan, M. Ünğan, & Z. Nedelko (Eds.), *Handbook of Research on Managerial Solutions in Non-Profit Organizations* (pp. 276–297). Hershey, PA: IGI Global. doi:10.4018/978-1-5225-0731-4.ch013

Fayezi, S., & Zomorrodi, M. (2016). Supply Chain Management: Developments, Theories and Models. In B. Christiansen (Ed.), *Handbook of Research on Global Supply Chain Management* (pp. 313–340). Hershey, PA: IGI Global. doi:10.4018/978-1-4666-9639-6.ch018

Fox, W. P. (2016). Multi-Attribute Decision Making in Risk Analysis. In D. Jakóbczak (Ed.), *Analyzing Risk through Probabilistic Modeling in Operations Research* (pp. 246–290). Hershey, PA: IGI Global. doi:10.4018/978-1-4666-9458-3.ch011

Fox, W. P. (2016). Perfect Partners of Mathematical Modeling with Technology in Risk Assessment. In D. Jakóbczak (Ed.), *Analyzing Risk through Probabilistic Modeling in Operations Research* (pp. 61–90). Hershey, PA: IGI Global. doi:10.4018/978-1-4666-9458-3.ch003

Giacomini, D., Chiaf, E., & Mazzoleni, M. B. (2017). How to Measure Performance in Cooperatives?: A Multiple Case Study. In V. Potocan, M. Ünğan, & Z. Nedelko (Eds.), *Handbook of Research on Managerial Solutions in Non-Profit Organizations* (pp. 343–361). Hershey, PA: IGI Global. doi:10.4018/978-1-5225-0731-4.ch016

Giri, B. C. (2016). Risk-Averse Newsboy Problem with Incomplete Demand Information: Risk-Averse Newsboy Problem. In D. Jakóbczak (Ed.), *Analyzing Risk through Probabilistic Modeling in Operations Research* (pp. 344–353). Hershey, PA: IGI Global. doi:10.4018/978-1-4666-9458-3.ch014

Glavee-Geo, R. (2016). The History and Development of Purchasing Management and Its Theoretical Framework: A Review of Transaction Cost Economics. In B. Christiansen (Ed.), *Handbook of Research on Global Supply Chain Management* (pp. 379–401). Hershey, PA: IGI Global. doi:10.4018/978-1-4666-9639-6.ch022

Gökgöz, F., & Atmaca, M. E. (2017). An Optimal Asset Allocation in Electricity Generation Market for the Policy Makers and Stakeholders. In V. Potocan, M. Ünğan, & Z. Nedelko (Eds.), *Handbook of Research on Managerial Solutions in Non-Profit Organizations* (pp. 448–482). Hershey, PA: IGI Global. doi:10.4018/978-1-5225-0731-4.ch021

Gökgöz, F., & Atmaca, M. E. (2017). Ranking Turkish Universities Based on Performance Evaluation via DEMATEL-AHP Approach. In V. Potocan, M. Ünğan, & Z. Nedelko (Eds.), *Handbook of Research on Managerial Solutions in Non-Profit Organizations* (pp. 421–447). Hershey, PA: IGI Global. doi:10.4018/978-1-5225-0731-4.ch020

Gomes, P., Camões, S. M., & Carvalho, J. (2016). Determinants of the Design and Use of PMS in Portuguese Government Agencies: A Complementary Theoretical Approach. In A. Ferreira, G. Azevedo, J. Oliveira, & R. Marques (Eds.), *Global Perspectives on Risk Management and Accounting in the Public Sector* (pp. 320–345). Hershey, PA: IGI Global. doi:10.4018/978-1-4666-9803-1.ch016

Gul, S., & Williams, T. (2017). Project Complexity for the Uninitiated. In Y. Raydugin (Ed.), *Handbook of Research on Leveraging Risk and Uncertainties for Effective Project Management* (pp. 202–219). Hershey, PA: IGI Global. doi:10.4018/978-1-5225-1790-0.ch010

Gumz, J. (2017). Managing Change: Strategies and Tactics to Review the Portfolio. In L. Romano (Ed.), *Project Portfolio Management Strategies for Effective Organizational Operations* (pp. 334–357). Hershey, PA: IGI Global. doi:10.4018/978-1-5225-2151-8.ch014

Hartlieb, S., & Silvius, G. (2017). Handling Uncertainty in Project Management and Business Development: Similarities and Differences. In Y. Raydugin (Ed.), *Handbook of Research on Leveraging Risk and Uncertainties for Effective Project Management* (pp. 337–362). Hershey, PA: IGI Global. doi:10.4018/978-1-5225-1790-0.ch016

Hass, K. B. (2017). Living on the Edge: Managing Project Complexity. In Y. Raydugin (Ed.), *Handbook of Research on Leveraging Risk and Uncertainties for Effective Project Management* (pp. 177–201). Hershey, PA: IGI Global. doi:10.4018/978-1-5225-1790-0.ch009

Hollmann, J. K. (2017). Complexity Risk and Modeling Disorder. In Y. Raydugin (Ed.), *Handbook of Research on Leveraging Risk and Uncertainties for Effective Project Management* (pp. 139–151). Hershey, PA: IGI Global. doi:10.4018/978-1-5225-1790-0.ch007

Hollmann, J. K. (2017). Systemic Risks and Parametric Modeling. In Y. Raydugin (Ed.), *Handbook of Research on Leveraging Risk and Uncertainties for Effective Project Management* (pp. 61–75). Hershey, PA: IGI Global. doi:10.4018/978-1-5225-1790-0.ch003

Hu, N., & Cheng, H. (2016). Survival Analysis and ROC Analysis in Analyzing Credit Risks: Assessing Default Risks Over Time. In D. Jakóbczak (Ed.), *Analyzing Risk through Probabilistic Modeling in Operations Research* (pp. 380–403). Hershey, PA: IGI Global. doi:10.4018/978-1-4666-9458-3.ch016

Huemann, M., & Ringhofer, C. (2016). Challenge or Potential? Risk Identification in the Context of Sustainable Development: A Case Study. In C. Bodea, A. Purnus, M. Huemann, & M. Hajdu (Eds.), *Managing Project Risks for Competitive Advantage in Changing Business Environments* (pp. 87–105). Hershey, PA: IGI Global. doi:10.4018/978-1-5225-0335-4.ch005

Hulett, D. T. (2017). Monte Carlo Simulation for Integrated Cost-Schedule Risk Analysis: Concepts, Methods, and Tools for Risk Analysis and Mitigation. In Y. Raydugin (Ed.), *Handbook of Research on Leveraging Risk and Uncertainties for Effective Project Management* (pp. 29–60). Hershey, PA: IGI Global. doi:10.4018/978-1-5225-1790-0.ch002

Ivančević, K. (2017). Insurance as an Optimization Tool for Risk Management in Small and Medium-Sized Enterprises. In M. Vemić (Ed.), *Optimal Management Strategies in Small and Medium Enterprises* (pp. 178–197). Hershey, PA: IGI Global. doi:10.4018/978-1-5225-1949-2.ch009

Jakóbczak, D. J. (2016). Data Extrapolation via Curve Modeling in Analyzing Risk: Value Anticipation for Decision Making. In D. Jakóbczak (Ed.), *Analyzing Risk through Probabilistic Modeling in Operations Research* (pp. 1–31). Hershey, PA: IGI Global. doi:10.4018/978-1-4666-9458-3.ch001

Jakóbczak, D. J. (2016). The Method of Probabilistic Nodes Combination in Decision Making and Risk Analysis: Data Extrapolation in Planning. In D. Jakóbczak (Ed.), *Analyzing Risk through Probabilistic Modeling in Operations Research* (pp. 32–60). Hershey, PA: IGI Global. doi:10.4018/978-1-4666-9458-3.ch002

Jarman, M. F., & Dreher, L. G. (2017). Missed Opportunities from Misalignment of Project and Operational Requirements. In Y. Raydugin (Ed.), *Handbook of Research on Leveraging Risk and Uncertainties for Effective Project Management* (pp. 363–383). Hershey, PA: IGI Global. doi:10.4018/978-1-5225-1790-0.ch017

Kač, S. M., Tamše, T., & Gorenak, I. (2017). Intercultural Education and Training of Slovene Police Officers for the Purpose of International Missions. In V. Potocan, M. Üngan, & Z. Nedelko (Eds.), *Handbook of Research on Managerial Solutions in Non-Profit Organizations* (pp. 298–320). Hershey, PA: IGI Global. doi:10.4018/978-1-5225-0731-4.ch014

Kafi, F., & Kafi, M. (2016). Information Quality in Supply Chain Software. In B. Christiansen (Ed.), *Handbook of Research on Global Supply Chain Management* (pp. 105–126). Hershey, PA: IGI Global. doi:10.4018/978-1-4666-9639-6.ch006

Kasemsap, K. (2016). Fostering Supply Chain Management in Global Business. In B. Christiansen (Ed.), *Handbook of Research on Global Supply Chain Management* (pp. 45–71). Hershey, PA: IGI Global. doi:10.4018/978-1-4666-9639-6.ch003

Kittilaksanawong, W. (2016). Value Creation and Appropriation in Buyer-Supplier Relationships: Governance, Competition and Cultures. In B. Christiansen (Ed.), *Handbook of Research on Global Supply Chain Management* (pp. 127–138). Hershey, PA: IGI Global. doi:10.4018/978-1-4666-9639-6.ch007

Klepac, G., Mrsic, L., & Kopal, R. (2016). Efficient Risk Profiling Using Bayesian Networks and Particle Swarm Optimization Algorithm. In D. Jakóbczak (Ed.), *Analyzing Risk through Probabilistic Modeling in Operations Research* (pp. 91–124). Hershey, PA: IGI Global. doi:10.4018/978-1-4666-9458-3.ch004

Kopp, U. (2016). Easier Identification of Risks and Uncertainties with Project Risk Constellations. In C. Bodea, A. Purnus, M. Huemann, & M. Hajdu (Eds.), *Managing Project Risks for Competitive Advantage in Changing Business Environments* (pp. 106–127). Hershey, PA: IGI Global. doi:10.4018/978-1-5225-0335-4.ch006

Kucukbay, F. (2017). Cash Management in Non-Profit Organizations. In V. Potocan, M. Üngan, & Z. Nedelko (Eds.), *Handbook of Research on Managerial Solutions in Non-Profit Organizations* (pp. 94–110). Hershey, PA: IGI Global. doi:10.4018/978-1-5225-0731-4.ch005

Kutsch, E., Hällgren, M., & Turner, N. (2016). The Internalization of Attention at 28,000 Feet: Revisiting the K2 2008 Disaster. In C. Bodea, A. Purnus, M. Huemann, & M. Hajdu (Eds.), *Managing Project Risks for Competitive Advantage in Changing Business Environments* (pp. 196–215). Hershey, PA: IGI Global. doi:10.4018/978-1-5225-0335-4.ch010

Lahmiri, S., & Gagnon, S. (2016). A Sequential Probabilistic System for Bankruptcy Data Classification. In D. Jakóbczak (Ed.), *Analyzing Risk through Probabilistic Modeling in Operations Research* (pp. 138–147). Hershey, PA: IGI Global. doi:10.4018/978-1-4666-9458-3.ch006

Lederer, M., Kurz, M., & Lazarov, P. (2017). Usage and Suitability of Methods for Strategic Business Process Initiatives: A Multi Case Study Research. *International Journal of Productivity Management and Assessment Technologies*, *5*(1), 40–51. doi:10.4018/IJPMAT.2017010103

Lee, I. (2017). A Social Enterprise Business Model and a Case Study of Pacific Community Ventures (PCV). In V. Potocan, M. Üngan, & Z. Nedelko (Eds.), *Handbook of Research on Managerial Solutions in Non-Profit Organizations* (pp. 182–204). Hershey, PA: IGI Global. doi:10.4018/978-1-5225-0731-4.ch009

Levin, G., & Pitotti, N. (2017). Program and Portfolio Management Relationship and Support. In L. Romano (Ed.), *Project Portfolio Management Strategies for Effective Organizational Operations* (pp. 310–333). Hershey, PA: IGI Global. doi:10.4018/978-1-5225-2151-8.ch013

Liu, Z. (2016). Structuring and Managing Supply Network: A Review of Current Literature and Conceptual Framework. In B. Christiansen (Ed.), *Handbook of Research on Global Supply Chain Management* (pp. 341–353). Hershey, PA: IGI Global. doi:10.4018/978-1-4666-9639-6.ch019

Lokuwaduge, C. S. (2016). Exploring the New Public Management (NPM)-Based Reforms in the Public Sector Accounting: A Sri Lankan Study. In A. Ferreira, G. Azevedo, J. Oliveira, & R. Marques (Eds.), *Global Perspectives on Risk Management and Accounting in the Public Sector* (pp. 49–67). Hershey, PA: IGI Global. doi:10.4018/978-1-4666-9803-1.ch003

Lorber, M., Treven, S., & Mumel, D. (2017). Leaders' Behavior in Association with Job Satisfaction and Organizational Commitment. In V. Potocan, M. Üngan, & Z. Nedelko (Eds.), *Handbook of Research on Managerial Solutions in Non-Profit Organizations* (pp. 111–133). Hershey, PA: IGI Global. doi:10.4018/978-1-5225-0731-4.ch006

Maciejczak, M., & Słodki, A. (2017). Initial Price Strategies of Polish Micro and Small Enterprises: An Application of Game Theory for Industrial Organization of the SME Sector. In M. Vemić (Ed.), *Optimal Management Strategies in Small and Medium Enterprises* (pp. 126–143). Hershey, PA: IGI Global. doi:10.4018/978-1-5225-1949-2.ch007

Magalhães, F. R., & Santos, C. (2016). Online Financial Transparency: Local Governments of the MERCOSUR Member Countries. In A. Ferreira, G. Azevedo, J. Oliveira, & R. Marques (Eds.), *Global Perspectives on Risk Management and Accounting in the Public Sector* (pp. 252–273). Hershey, PA: IGI Global. doi:10.4018/978-1-4666-9803-1.ch013

Malega, P. (2017). Small and Medium Enterprises in the Slovak Republic: Status and Competitiveness of SMEs in the Global Markets and Possibilities of Optimization. In M. Vemić (Ed.), *Optimal Management Strategies in Small and Medium Enterprises* (pp. 102–124). Hershey, PA: IGI Global. doi:10.4018/978-1-5225-1949-2.ch006

Malek, N. A., & Takala, J. A. (2017). Regression Analysis for Environmental Practices: Participation among Finnish SMEs. In M. Vemić (Ed.), *Optimal Management Strategies in Small and Medium Enterprises* (pp. 44–59). Hershey, PA: IGI Global. doi:10.4018/978-1-5225-1949-2.ch003

Malewska, K. M. (2017). Intuition in Decision-Making on the Example of a Non-Profit Organization. In V. Potocan, M. Üngan, & Z. Nedelko (Eds.), *Handbook of Research on Managerial Solutions in Non-Profit Organizations* (pp. 378–399). Hershey, PA: IGI Global. doi:10.4018/978-1-5225-0731-4.ch018

Malik, M. S. (2017). Risk Based Contingency in Project Cost Estimates. In Y. Raydugin (Ed.), *Handbook of Research on Leveraging Risk and Uncertainties for Effective Project Management* (pp. 130–138). Hershey, PA: IGI Global. doi:10.4018/978-1-5225-1790-0.ch006

Mendes, H. C., Santos, C., Ferreira, A. D., Marques, R. P., Azevedo, G. M., & Oliveira, J. D. (2016). Local Authorities and the Disclosure of Financial Information via the Internet: The Portuguese Case. In A. Ferreira, G. Azevedo, J. Oliveira, & R. Marques (Eds.), *Global Perspectives on Risk Management and Accounting in the Public Sector* (pp. 274–297). Hershey, PA: IGI Global. doi:10.4018/978-1-4666-9803-1.ch014

Miller, T., de Matta, R., & Xu, M. (2016). High Level Inventory Network Modeling Approaches. In B. Christiansen (Ed.), *Handbook of Research on Global Supply Chain Management* (pp. 15–44). Hershey, PA: IGI Global. doi:10.4018/978-1-4666-9639-6.ch002

Milošević, O. (2017). Critical Review of SME Regulation Optimization in Serbia: A Reflection on Harmonization with the EU Acquis. In M. Vemić (Ed.), *Optimal Management Strategies in Small and Medium Enterprises* (pp. 79–101). Hershey, PA: IGI Global. doi:10.4018/978-1-5225-1949-2.ch005

Mogos, R., Bodea, C., Stancu, S., Purnus, A., & Dascalu, M. (2016). Educating Project Managers to Deal with Project Risks: Improvement of the Educational Programmes Design by Using Curriculum Management Systems. In C. Bodea, A. Purnus, M. Huemann, & M. Hajdu (Eds.), *Managing Project Risks for Competitive Advantage in Changing Business Environments* (pp. 231–257). Hershey, PA: IGI Global. doi:10.4018/978-1-5225-0335-4.ch012

Molnar, R. M. (2017). Structural Change Management for Sustainable SME Development: Applying Classical Management Tools. In M. Vemić (Ed.), *Optimal Management Strategies in Small and Medium Enterprises* (pp. 348–369). Hershey, PA: IGI Global. doi:10.4018/978-1-5225-1949-2.ch016

Nandonde, F. A., & Nguni, W. (2016). Relationship Establishment in SCM in a Market with Enforcement and Regulation Challenges: Case of Tanzania. In B. Christiansen (Ed.), *Handbook of Research on Global Supply Chain Management* (pp. 354–365). Hershey, PA: IGI Global. doi:10.4018/978-1-4666-9639-6.ch020

Nedelko, Z., & Potocan, V. (2017). Management Solutions in Non-Profit Organizations: Case of Slovenia. In V. Potocan, M. Üngan, & Z. Nedelko (Eds.), *Handbook of Research on Managerial Solutions in Non-Profit Organizations* (pp. 1–22). Hershey, PA: IGI Global. doi:10.4018/978-1-5225-0731-4.ch001

Nersesian, R. L. (2017). Incorporating Simulation in Project Scoping. In Y. Raydugin (Ed.), *Handbook of Research on Leveraging Risk and Uncertainties for Effective Project Management* (pp. 406–434). Hershey, PA: IGI Global. doi:10.4018/978-1-5225-1790-0.ch019

Nesbitt, R., & Kotb, A. (2016). Governance in NHS Foundation Trusts: Insights from Company Secretaries. In A. Ferreira, G. Azevedo, J. Oliveira, & R. Marques (Eds.), *Global Perspectives on Risk Management and Accounting in the Public Sector* (pp. 167–187). Hershey, PA: IGI Global. doi:10.4018/978-1-4666-9803-1.ch009

Nonino, F. (2017). Project Selection Frameworks and Methodologies for Reducing Risks in Project Portfolio Management. In L. Romano (Ed.), *Project Portfolio Management Strategies for Effective Organizational Operations* (pp. 245–263). Hershey, PA: IGI Global. doi:10.4018/978-1-5225-2151-8.ch010

Olson, B. (2017). Optimizing Portfolio Value through Comprehensive Project Metrics. In L. Romano (Ed.), *Project Portfolio Management Strategies for Effective Organizational Operations* (pp. 178–201). Hershey, PA: IGI Global. doi:10.4018/978-1-5225-2151-8.ch007

Parth, F. R. (2017). Preparing the Organization for Portfolio Management: Overcoming Resistance and Obstacles. In L. Romano (Ed.), *Project Portfolio Management Strategies for Effective Organizational Operations* (pp. 119–152). Hershey, PA: IGI Global. doi:10.4018/978-1-5225-2151-8.ch005

Parth, F. R. (2017). Project Portfolio Management Growth and Operation: Portfolio Management Structure, Operations, Risk, and Growth. In L. Romano (Ed.), *Project Portfolio Management Strategies for Effective Organizational Operations* (pp. 264–287). Hershey, PA: IGI Global. doi:10.4018/978-1-5225-2151-8.ch011

Pečarič, M. (2016). The Awareness of Mentality in Public Administration as the Key for the Management of Its Complexity. In A. Ferreira, G. Azevedo, J. Oliveira, & R. Marques (Eds.), *Global Perspectives on Risk Management and Accounting in the Public Sector* (pp. 1–24). Hershey, PA: IGI Global. doi:10.4018/978-1-4666-9803-1.ch001

Peng, H. (2016). Condition-Based Maintenance: Capital Goods Industry. In D. Jakóbczak (Ed.), *Analyzing Risk through Probabilistic Modeling in Operations Research* (pp. 292–320). Hershey, PA: IGI Global. doi:10.4018/978-1-4666-9458-3.ch012

Pereira de Campos, C. M., Rodrigues, L. L., & Jorge, S. M. (2016). The Role of Management Accounting Systems in Public Hospitals and the Construction of Budgets: A Literature Review. In A. Ferreira, G. Azevedo, J. Oliveira, & R. Marques (Eds.), *Global Perspectives on Risk Management and Accounting in the Public Sector* (pp. 366–389). Hershey, PA: IGI Global. doi:10.4018/978-1-4666-9803-1.ch018

Pham, T. S., Darabi, F., & Wilmot, N. V. (2016). International Supply Chain Case Study. In B. Christiansen (Ed.), *Handbook of Research on Global Supply Chain Management* (pp. 205–226). Hershey, PA: IGI Global. doi:10.4018/978-1-4666-9639-6.ch012

Pinho, M., & Santos, C. (2016). Plan for Prevention of Risks of Corruption and Related Infractions: The Application of FMEA Methodology. In A. Ferreira, G. Azevedo, J. Oliveira, & R. Marques (Eds.), *Global Perspectives on Risk Management and Accounting in the Public Sector* (pp. 390–412). Hershey, PA: IGI Global. doi:10.4018/978-1-4666-9803-1.ch019

Plumery, R. C. (2017). Risk-Centric Performance Measurement: Deriving Best Value from Performance Management Systems. In Y. Raydugin (Ed.), *Handbook of Research on Leveraging Risk and Uncertainties for Effective Project Management* (pp. 318–336). Hershey, PA: IGI Global. doi:10.4018/978-1-5225-1790-0.ch015

Postăvaru, N., & Leonte, B. (2016). Project Portfolio Risk Management: Managing Risk in Case of Investment Portfolio. In C. Bodea, A. Purnus, M. Huemann, & M. Hajdu (Eds.), *Managing Project Risks for Competitive Advantage in Changing Business Environments* (pp. 279–300). Hershey, PA: IGI Global. doi:10.4018/978-1-5225-0335-4.ch014

Qi, A., & Zheng, L. (2016). Project Risk Management: A Chinese Perspective. In C. Bodea, A. Purnus, M. Huemann, & M. Hajdu (Eds.), *Managing Project Risks for Competitive Advantage in Changing Business Environments* (pp. 45–69). Hershey, PA: IGI Global. doi:10.4018/978-1-5225-0335-4.ch003

Raue, S., & Klein, L. (2016). Systemic Risk Management: A Practice Approach to the Systemic Management of Project Risk. In C. Bodea, A. Purnus, M. Huemann, & M. Hajdu (Eds.), *Managing Project Risks for Competitive Advantage in Changing Business Environments* (pp. 70–85). Hershey, PA: IGI Global. doi:10.4018/978-1-5225-0335-4.ch004

Raydugin, Y. (2017). Integrated Risk-Based and Economic-Based (IRBEB) Methodology for Selection of Project Alternatives. In Y. Raydugin (Ed.), *Handbook of Research on Leveraging Risk and Uncertainties for Effective Project Management* (pp. 459–469). Hershey, PA: IGI Global. doi:10.4018/978-1-5225-1790-0.ch021

Raydugin, Y. (2017). Project Risk Management: Popular Fallacies and Overlooked Best Practices. In Y. Raydugin (Ed.), *Handbook of Research on Leveraging Risk and Uncertainties for Effective Project Management* (pp. 1–28). Hershey, PA: IGI Global. doi:10.4018/978-1-5225-1790-0.ch001

Rinderu, P., Voiculescu, C. I., & Visanescu, D. L. (2016). Risks in Implementing EU-Funded Projects in Romanian Public Higher Education System: Measures for Better Institutional Response. In A. Ferreira, G. Azevedo, J. Oliveira, & R. Marques (Eds.), *Global Perspectives on Risk Management and Accounting in the Public Sector* (pp. 413–441). Hershey, PA: IGI Global. doi:10.4018/978-1-4666-9803-1.ch020

Romano, L. (2017). Adaptive Portfolio Management. In L. Romano (Ed.), *Project Portfolio Management Strategies for Effective Organizational Operations* (pp. 153–177). Hershey, PA: IGI Global. doi:10.4018/978-1-5225-2151-8.ch006

Romano, L. (2017). Portfolio Management as a Step into the Future. In L. Romano (Ed.), *Project Portfolio Management Strategies for Effective Organizational Operations* (pp. 1–38). Hershey, PA: IGI Global. doi:10.4018/978-1-5225-2151-8.ch001

Romano, L., Grimaldi, R., & Colasuonno, F. S. (2017). Demand Management as a Success Factor in Project Portfolio Management. In L. Romano (Ed.), *Project Portfolio Management Strategies for Effective Organizational Operations* (pp. 202–219). Hershey, PA: IGI Global. doi:10.4018/978-1-5225-2151-8.ch008

Rostek, K. B. (2016). Risk Management: Role and Importance in Business Organization. In D. Jakóbczak (Ed.), *Analyzing Risk through Probabilistic Modeling in Operations Research* (pp. 149–178). Hershey, PA: IGI Global. doi:10.4018/978-1-4666-9458-3.ch007

Sá, F., Rocha, Á., & Pérez-Cota, M. (2016). A Literature Review on Quality Models for Online E-Government Services. In A. Ferreira, G. Azevedo, J. Oliveira, & R. Marques (Eds.), *Global Perspectives on Risk Management and Accounting in the Public Sector* (pp. 151–166). Hershey, PA: IGI Global. doi:10.4018/978-1-4666-9803-1.ch008

Sanchez-Ruiz, L., & Blanco, B. (2017). Process Management for SMEs: Barriers, Enablers, and Benefits. In M. Vemić (Ed.), *Optimal Management Strategies in Small and Medium Enterprises* (pp. 293–319). Hershey, PA: IGI Global. doi:10.4018/978-1-5225-1949-2.ch014

Schoper, Y., Böhle, F., & Heidling, E. (2016). Coping Better with the Project's Unknown Unknowns: New Competences for Overcoming Uncertainty in Projects. In C. Bodea, A. Purnus, M. Huemann, & M. Hajdu (Eds.), *Managing Project Risks for Competitive Advantage in Changing Business Environments* (pp. 1–22). Hershey, PA: IGI Global. doi:10.4018/978-1-5225-0335-4.ch001

Serpell, A. F., Ferrada, X., & Rubio, L. (2016). Using a Knowledge-Based Approach to Foster the Use of Risk Management in Construction. In C. Bodea, A. Purnus, M. Huemann, & M. Hajdu (Eds.), *Managing Project Risks for Competitive Advantage in Changing Business Environments* (pp. 258–278). Hershey, PA: IGI Global. doi:10.4018/978-1-5225-0335-4.ch013

Shang, K. (2017). Improving Objectivity in Project Risk Management. In Y. Raydugin (Ed.), *Handbook of Research on Leveraging Risk and Uncertainties for Effective Project Management* (pp. 248–265). Hershey, PA: IGI Global. doi:10.4018/978-1-5225-1790-0.ch012

Sharma, S. K., Bhat, A., Kumar, V., & Agarwal, A. (2017). Path Analysis Model for Supply Chain Risk Management. *International Journal of Information Systems and Supply Chain Management, 10*(2), 21–41. doi:10.4018/IJISSCM.2017040102

Shih, C. P., & Orochena, O. D. (2016). Analyzing the Effect of Transformational Leadership on Innovation and Organizational Performance. *International Journal of Productivity Management and Assessment Technologies, 4*(2), 11–27. doi:10.4018/IJPMAT.2016070102

Silvius, G. (2016). Integrating Sustainability into Project Risk Management. In C. Bodea, A. Purnus, M. Huemann, & M. Hajdu (Eds.), *Managing Project Risks for Competitive Advantage in Changing Business Environments* (pp. 23–44). Hershey, PA: IGI Global. doi:10.4018/978-1-5225-0335-4.ch002

Soniewicki, M. (2017). The Importance of Market Orientation in Creating a Competitive Advantage of Micro, Small, and Medium-Sized Companies in the Internationalization Process. In M. Vemić (Ed.), *Optimal Management Strategies in Small and Medium Enterprises* (pp. 1–21). Hershey, PA: IGI Global. doi:10.4018/978-1-5225-1949-2.ch001

Sönmez, R. (2017). Value Creation through Social Alliances: Theoretical Considerations in Partnership Relationships. In V. Potocan, M. Üngan, & Z. Nedelko (Eds.), *Handbook of Research on Managerial Solutions in Non-Profit Organizations* (pp. 205–231). Hershey, PA: IGI Global. doi:10.4018/978-1-5225-0731-4.ch010

Sowa, F., & Staples, R. (2016). Public Administration in the Era of Late Neo-Liberalism: Placement Professionals and the NPM Regime. In A. Ferreira, G. Azevedo, J. Oliveira, & R. Marques (Eds.), *Global Perspectives on Risk Management and Accounting in the Public Sector* (pp. 25–48). Hershey, PA: IGI Global. doi:10.4018/978-1-4666-9803-1.ch002

Stancu, S. (2016). Managing Risks in the Innovation Projects in the Services Sector: A Case Study for the Mobile Communication. In C. Bodea, A. Purnus, M. Huemann, & M. Hajdu (Eds.), *Managing Project Risks for Competitive Advantage in Changing Business Environments* (pp. 174–195). Hershey, PA: IGI Global. doi:10.4018/978-1-5225-0335-4.ch009

Stefanova, J. S., & Wenner, Z. (2017). Entrepreneurship and Enterprise Value Creation in Support of Smart, Sustainable, and Inclusive Growth in the European Union. In M. Vemić (Ed.), *Optimal Management Strategies in Small and Medium Enterprises* (pp. 268–292). Hershey, PA: IGI Global. doi:10.4018/978-1-5225-1949-2.ch013

Stoian, M., & Stoian, R. (2016). Financial Risk Management: An Introduction. In C. Bodea, A. Purnus, M. Huemann, & M. Hajdu (Eds.), *Managing Project Risks for Competitive Advantage in Changing Business Environments* (pp. 301–311). Hershey, PA: IGI Global. doi:10.4018/978-1-5225-0335-4.ch015

Tavares, M. D., & Rodrigues, L. L. (2016). The Determinants of Sustainability Reporting of the Portuguese Public Sector Entities. In A. Ferreira, G. Azevedo, J. Oliveira, & R. Marques (Eds.), *Global Perspectives on Risk Management and Accounting in the Public Sector* (pp. 226–251). Hershey, PA: IGI Global. doi:10.4018/978-1-4666-9803-1.ch012

Timilsina, B. (2017). Overcoming the Barriers of Strategic Planning, Implementation, and Monitoring in Turbulent Business Environment: A Qualitative Study on Finnish SMEs. In M. Vemić (Ed.), *Optimal Management Strategies in Small and Medium Enterprises* (pp. 226–248). Hershey, PA: IGI Global. doi:10.4018/978-1-5225-1949-2.ch011

Tongiani, M. G., & Zhao, S. (2017). Traditional Italian Food Products on the Chinese Market: Best Practices for Italian Small and Medium Enterprises. In M. Vemić (Ed.), *Optimal Management Strategies in Small and Medium Enterprises* (pp. 22–43). Hershey, PA: IGI Global. doi:10.4018/978-1-5225-1949-2.ch002

Toth, T., & Sebestyen, Z. (2016). Project Risk Management Process for Professionals: A Value-Based Approach. In C. Bodea, A. Purnus, M. Huemann, & M. Hajdu (Eds.), *Managing Project Risks for Competitive Advantage in Changing Business Environments* (pp. 128–149). Hershey, PA: IGI Global. doi:10.4018/978-1-5225-0335-4.ch007

Townley, T. W. (2017). Modeling Risk and Opportunity Reality. In Y. Raydugin (Ed.), *Handbook of Research on Leveraging Risk and Uncertainties for Effective Project Management* (pp. 435–458). Hershey, PA: IGI Global. doi:10.4018/978-1-5225-1790-0.ch020

Trickey, G. (2017). Getting to Grips with Behavioural Bias: How Projects Fail and What To Do About It. In Y. Raydugin (Ed.), *Handbook of Research on Leveraging Risk and Uncertainties for Effective Project Management* (pp. 220–247). Hershey, PA: IGI Global. doi:10.4018/978-1-5225-1790-0.ch011

Tudor, C., & Tudor, M. (2016). On Using Monte Carlo Simulations for Project Risk Management. In C. Bodea, A. Purnus, M. Huemann, & M. Hajdu (Eds.), *Managing Project Risks for Competitive Advantage in Changing Business Environments* (pp. 150–173). Hershey, PA: IGI Global. doi:10.4018/978-1-5225-0335-4.ch008

Turner, R. (2016). Risks in Project Marketing. In C. Bodea, A. Purnus, M. Huemann, & M. Hajdu (Eds.), *Managing Project Risks for Competitive Advantage in Changing Business Environments* (pp. 217–230). Hershey, PA: IGI Global. doi:10.4018/978-1-5225-0335-4.ch011

Vaiappuri, S. K., Kamarulzaman, N. H., Vijayan, G., & Mukherjee, A. (2016). An Empirical View of Knowledge Management. In B. Christiansen (Ed.), *Handbook of Research on Global Supply Chain Management* (pp. 452–465). Hershey, PA: IGI Global. doi:10.4018/978-1-4666-9639-6.ch025

Vecchi, A., & Vallisi, V. (2016). Supply Chain Resilience. In B. Christiansen (Ed.), *Handbook of Research on Global Supply Chain Management* (pp. 147–160). Hershey, PA: IGI Global. doi:10.4018/978-1-4666-9639-6.ch009

Vemić, M. B. (2017). A Further Look at Working Capital Optimization in Medium-Sized Firms: Concepts and Evidence. In M. Vemić (Ed.), *Optimal Management Strategies in Small and Medium Enterprises* (pp. 144–177). Hershey, PA: IGI Global. doi:10.4018/978-1-5225-1949-2.ch008

Vemić, M. B. (2017). Financial Innovation in Medium-Sized Enterprises Optimizes Their Gravitation Towards Capital Markets: Financial Future in Perspective. In M. Vemić (Ed.), *Optimal Management Strategies in Small and Medium Enterprises* (pp. 198–224). Hershey, PA: IGI Global. doi:10.4018/978-1-5225-1949-2.ch010

Vijayan, G., Kamarulzaman, N. H., Mukherjee, A., & Vaiappuri, S. K. (2016). Strategic Value Creation in a Supply Chain. In B. Christiansen (Ed.), *Handbook of Research on Global Supply Chain Management* (pp. 186–204). Hershey, PA: IGI Global. doi:10.4018/978-1-4666-9639-6.ch011

Wodecka-Hyjek, A. (2017). Co-Operation between the Public Administration and Non-Profit Organisations as a Condition of the Development of Public Entrepreneurship: On the Example of the Selected World Solutions. In V. Potocan, M. Üngan, & Z. Nedelko (Eds.), *Handbook of Research on Managerial Solutions in Non-Profit Organizations* (pp. 253–275). Hershey, PA: IGI Global. doi:10.4018/978-1-5225-0731-4.ch012

Yordanova, S. D. (2017). Optimizing Virtual Communities in Tourism to Facilitate Development of Small and Medium-Sized Enterprises. In M. Vemić (Ed.), *Optimal Management Strategies in Small and Medium Enterprises* (pp. 60–78). Hershey, PA: IGI Global. doi:10.4018/978-1-5225-1949-2.ch004

Yumurtacı, I. Ö., & Oflaç, B. S. (2016). Supply Chain Management Practices in Retail-The Trends and Future Perspective. In B. Christiansen (Ed.), *Handbook of Research on Global Supply Chain Management* (pp. 366–378). Hershey, PA: IGI Global. doi:10.4018/978-1-4666-9639-6.ch021

Zhang, Y. (2016). An Empirical Investigation of the Role of E-Communication in International Collaborations. In B. Christiansen (Ed.), *Handbook of Research on Global Supply Chain Management* (pp. 85–104). Hershey, PA: IGI Global. doi:10.4018/978-1-4666-9639-6.ch005

Zomorrodi, M. (2016). Supply Chain Risk Management: A Review of the Literature. In B. Christiansen (Ed.), *Handbook of Research on Global Supply Chain Management* (pp. 516–530). Hershey, PA: IGI Global. doi:10.4018/978-1-4666-9639-6.ch029

About the Authors

Edem G. Tetteh is the founding Dean of Science, Technology, Engineering, and Mathematics (STEM) Division at Rowan College at Burlington County. Previously, he was interim Dean of Academic Affairs at Potomac State College of West Virginia University. He has held the position of Assistant Vice President for Academic Affairs and Associate Professor of Science, Technology, and Math at Paine College. He received both his BS in Manufacturing Systems and MS in Industrial Engineering from North Carolina Agricultural and Technical State University. He received his PhD in Technology from Purdue University-West Lafayette. Dr. Tetteh has authored many books and publications in the area of Ergonomics & Human Factors and Logistic & Supply Chain. Dr. Tetteh is a Lean Six Sigma Black Belt certified and has received several awards from the program.

Hans Chapman is the Program Director for the Associate in Engineering (AE) College Transfer program at Cape Fear Community College. Previously, he was an Associate Professor of Mechanical and Manufacturing at the School of Engineering and Information Systems (SEIS) at Morehead State University (MSU) in Morehead KY. Dr. Chapman obtained his PhD degree from the School of Mechanical and Materials Engineering at the University of Surrey, England. Dr. Chapman's research activities are in the area of Renewable Energy Systems Quality Engineering. He is also a reviewer for the International Journal of Modern, Mechanical and Manufacturing Systems, and Engineering (IJME). Dr. Chapman has been an Engineering faculty in the Kentucky Governor's Scholar Program (GSP). In his spare time, Dr. Chapman enjoys star-gazing, fishing, traveling with his family and sports.

Index

S

T

W

www.ingramcontent.com/pod-product-compliance
Lightning Source LLC
Chambersburg PA
CBHW080242211225
37091CB00003B/523

* 9 7 8 1 5 2 2 5 4 0 6 2 5 *